GREAT TRACTORS

FIELD-MARSHA

GREAT TRACTORS

MICHAEL WILLIAMS

Photography
Andrew Morland

BLANDFORD PRESS
Poole Dorset

First published in the U.K. 1982 by Blandford Press,
Link House, West Street, Poole, Dorset, BH15 1LL.

Reprinted 1983
Reprinted 1984
Reprinted 1986

Distributed in Australia
by Capricorn Link (Australia) Pty Ltd,
PO Box 665, Lane Cove,
NSW 2066

Distributed in North America by
Diamond Farm Book Publishers
RR3
Brighton
Ontario
KOK 1HO Canada

Box 537
Alexandria Bay
New York 13607
USA

British Library Cataloguing in Publication Data

Williams, Michael
Great tractors.
1. Tractors – History
I. Title
629.2′25 TL233

ISBN 0 7137 1205 8

Typeset by Asco Trade Typesetting Ltd,
Hong Kong

Printed by South China Printing Co., Hong Kong

CONTENTS

ACKNOWLEDGEMENTS

The author and publishers would like to thank all those who kindly allowed their tractors to be photographed for this book, including especially Messrs Boulters of Banwell; C.H. Doble (The Cletrac Collection); Philip Jenkinson of Shebbear Farm Museum and the Ontario Agricultural Museum.

We would also like to thank all of the following who provided photographs which have been included in this book: Allis-Chalmers; British Leyland Tractors; Caterpillar Tractor Company; County Commercial Cars; David Brown Tractors; Deere and Company; S. Dickinson; Ernest Doe and Sons; Fiat Trattori; Ford Motor Company; Greenfield Village and Henry Ford Museum; Howard Rotavator Company; Ray Hooley; Hurlimann Traktoren; Hungarian National Museum of Agriculture; International Harvester Company; Gerald Lambert; Landini; E. Liebl; Henry Marshall; Massey-Ferguson; Museum of English Rural Life; National Institute of Agricultural Engineering; National Museum of Antiquities, Scotland; Ontario Agricultural Museum; N. Pross; Ransomes, Sims and Jefferies; Renault; J. Saunderson; Smithsonian Institution; Trans-Antarctic Expedition; University of Reading; Vickers Ltd; Western Development Museum; Wye College, London University.

AUTHOR'S NOTE

The modern farm tractor is a highly versatile power source and has become an essential factor in agricultural mechanisation to provide the food for a hungry world. It is the result of nearly a century of development, reflecting the contributions of many inventors and engineers.

During the development of the tractor, a great many makes and models have been produced in a remarkable range of shapes and sizes. This book describes a small selection of these tractors – some which were successes and some which failed, some which were important landmarks in tractor history and some which were hardly noticed.

This selection represents just a few examples of tractor development from over 90 years of progress. Perhaps some of the omissions may be included in a further volume, but meanwhile I hope this book contains something of interest to anyone who likes old tractors.

Michael Williams

CASE EXPERIMENTAL PROTOTYPE

The first Case tractor was completed in 1892. The J.I. Case Threshing Machine Company was then heavily involved in the agricultural steam-engine business and, a few years later, became the largest manufacturer of portable and traction engines in the world. With such a big stake in the steam business it was surprising that Case took such an early interest in the petrol engine. Most of the big steam-engine companies in the USA and Europe still ignored the internal combustion engine and the tractor.

The word 'tractor' had not been coined in 1892 and would not have been entirely appropriate for most of the experimental machines built before the end of the century. With very few exceptions, these were more like steam traction engines in size and weight and were intended to do similar work. There was little attempt to design a tractor with versatility and lightness to compete with the farm horse.

Traction-engine influence is clearly evident in the Case design. It was built on the wheels and frame of a traction engine, with typical traction-engine steering gear. The power unit was a

The first Case tractor.

'balanced' engine designed by William Patterson, with two horizontal cylinders which were opposed. The ignition system was a particularly crude form of make-and-break and the carburettor worked on the notoriously inaccurate surface principle and was the large horizontal cylinder at the front of the tractor.

There were single speeds forward and reverse, with exposed-transmission gearing and a sliding key to select the required gear or neutral.

Development work continued, apparently, for several years while the company tried to overcome serious problems in the engine design. As the fuel and ignition systems were basically unreliable, it is unlikely that much real progress was made. The project was abandoned and this was probably a sensible decision. Internal combustion engines in the 1890s were not at an acceptable stage of development for farm conditions, while the steam engine had already achieved a reasonable degree of reliability.

The company showed equally good commercial judgment almost 20 years later when they decided to move back into the tractor business. Steam on the farm had reached its peak and the tractor was poised on the edge of a tremendous period of expansion in the USA.

Case became one of the most successful companies in the tractor business and carried the eagle trademark to many parts of the world. The Case eagle first appeared on the company's threshing machines in 1865. It was based on 'Old Abe', the eagle which had been a regimental mascot in the American Civil War and had been named after Abraham Lincoln. Then the Case eagle was perched on a branch, but, in 1894, the emblem was redesigned to place the eagle on top of the world. This was to become perhaps the most distinctive trademark in the farm equipment business for 75 years.

The 'Old Abe' trademark photographed on a 1939 tractor.

HORNSBY-AKROYD

Richard Hornsby & Sons achieved a remarkable list of distinctions as pioneers of tractor development in Great Britain, but the commercial rewards for their efforts must have been a bitter disappointment.

The first Hornsby tractor was completed at the company's factory at Grantham, Lincolnshire, in 1896. It was known as the Hornsby-Akroyd Patent Safety Oil Traction Engine and weighed $8\frac{1}{2}$ tons.

At that time the Hornsby company held the manufacturing rights to the Stuart and Binney engine design. This was an oil-burning engine which had achieved a good reputation for reliability. A feature of the design was the starting system. This consisted of a blow-lamp, which was used to create a hot-spot inside the cylinder head. This avoided the need for any form of electric ignition.

This was the type of engine used by Hornsby for their tractor. Four engine sizes were available, from 16 to 32 hp, but only 20-hp engines were used in the four tractors which were manufactured. The 20-hp engine was a four-stroke, single-cylinder, and was horizontal and water-cooled.

In 1897, the Hornsby became the first tractor to appear at the Royal Show and the first to be awarded the prestigious Silver Medal of the Royal Agricultural Society of England. In September of the same year, Mr Locke-King, a Surrey landowner, bought the 1896 tractor – the first time a tractor had been sold in Great Britain.

Three more agricultural tractors were built to the same basic design and all of these were exported to Australia, where one of them still survives.

In 1903, the Hornsby company began a determined effort to do business with the British War Office. A Military tractor powered by a twin-cylinder engine was built to replace steam traction engines for heavy, long-distance transport. The British Army was interested in buying equipment of this kind and the Hornsby tractor performed

Hornsby-Akroyd tractor of 1896.

sufficiently well in trials to win a £1000 prize, which was awarded by the War Office, but the big orders failed to materialise.

The following year, the original 1896 tractor was returned to the works from Surrey. Hornsby's chief engineer, David Roberts, had recently taken out a patent for crawler tracks and the old tractor was used as a test-bed and became the first successful tracklayer with a track-steering mechanism.

This tracklayer, and others built by the company, performed in a series of trials designed to impress the War Office with the military potential of a crawler tractor capable of travelling through mud and over barbed-wire entanglements. Hornsby had developed a predecessor of the army tank, but there was little response from the British Army.

Just before World War 1 brought the need for a tracklaying army vehicle into prominence, the Hornsby company sold their crawler-track patents to the Holt company in the USA. In 1916, the first British Army tanks were rolling towards the German trenches in the Somme battlefield to start one of the biggest revolutions in military history.

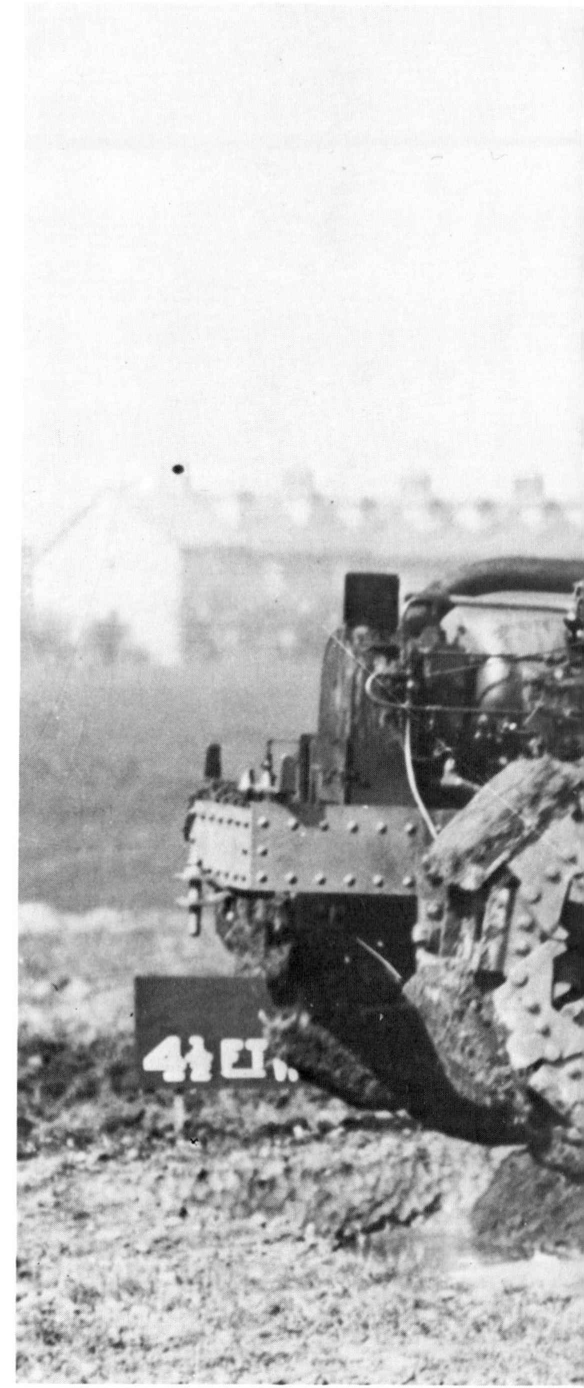

The Hornsby tractor converted to crawler tracks.

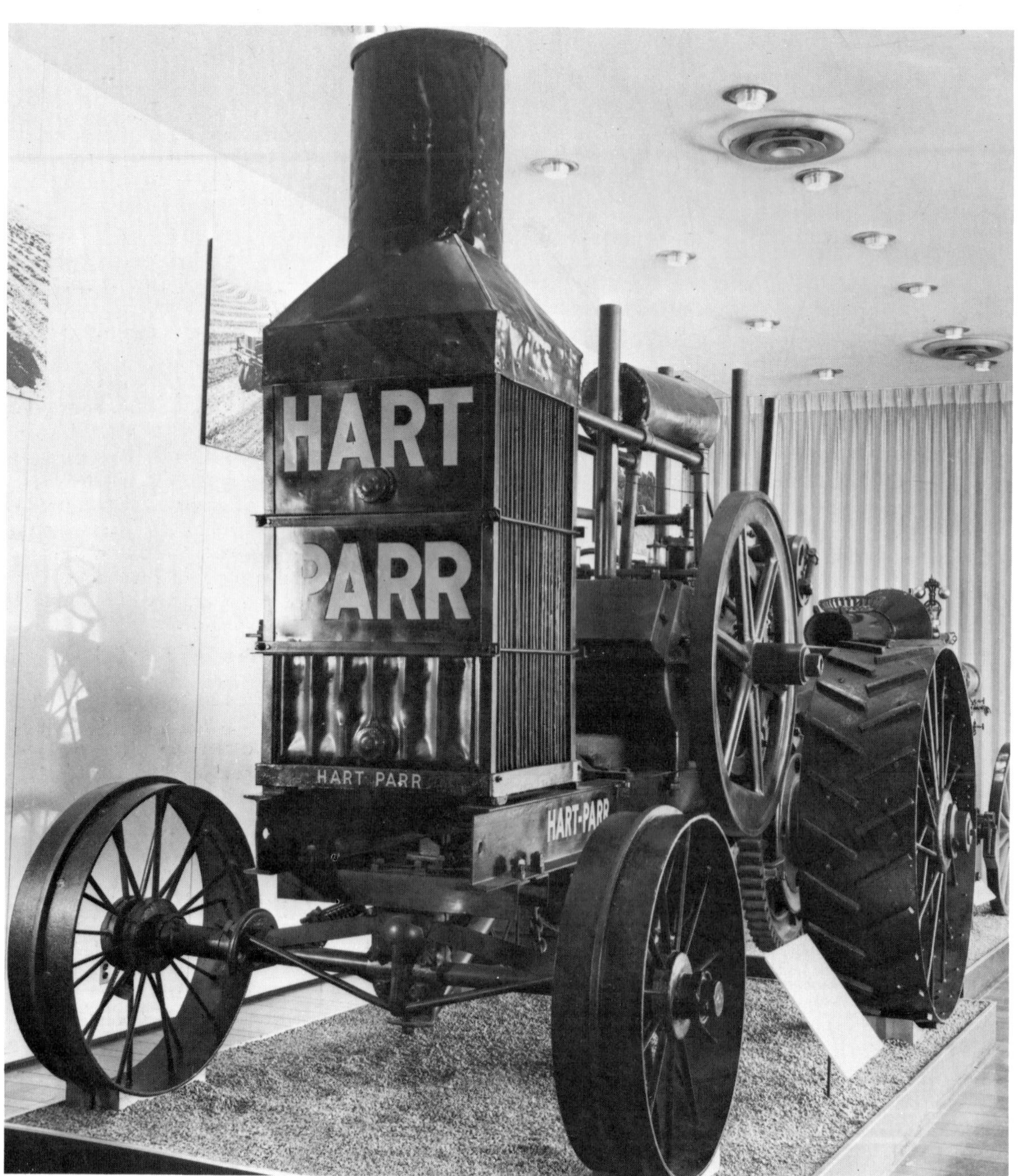

Hart-Parr No. 3 in its resting place in the Smithsonian Institution.

HART-PARR 18-30

Charles Hart and Charles Parr, who formed the Hart-Parr company, were engineering students together at the University of Wisconsin. Both had a special interest in engine design and, after graduating in 1896, they formed a company at Madison, Wisconsin, to build stationary engines. One of the features of their engine design was the use of oil instead of water for cooling. This helped establish the engines in areas where there was a risk of frost damage.

As the engine business expanded, Hart and Parr decided, in 1901, to move the company out of Madison to Charles City, Iowa, where Charles Hart had spent his childhood. With more factory space available, work started on the first Hart-Parr tractor and this was sold, in 1902, to an Iowa farmer. This tractor, which was claimed to have given years of reliable service, included many features which were to become typical of Hart-Parr design for many years. One of these features was the oil-cooling system and others were the heavy, rugged construction, the twin-cylinder engine and the steel-girder chassis.

The 18–30 tractor was completed in 1903 and was the third design produced by the Hart-Parr partnership. It introduced the massive rectangular cooling tower which remained a feature of nearly all Hart-Parr tractors for more than 15 years. The cooling system circulated hot oil from the engine through vertical tubes in the tower. An air flow inside the tower was encouraged by directing exhaust gases from the engine to create an upward movement through ducts between the radiator sections.

The power unit was rated at 30 hp and was a horizontal design with two large cylinders, each of approximately 10-in bore and 13-in stroke. The engine was designed to operate at up to about 300 rpm and burned petrol with a water-injection system to prevent pre-ignition.

Instead of sparking plugs, the Hart-Parr engine used eight Edison sealed wet cells and a make-and-break ignition system. The lubrication system was a simple arrangement of sight-feed oilers, which were probably quite inadequate for the exposed bearings when working in dusty conditions.

The 18–30 tractor was the forerunner of the first Hart-Parr tractor to be produced in commercial numbers. This was the 17–30, of which 15 are claimed to have been built in 1903. The most obvious difference between the 18–30 and the 17–30s which followed it, was the addition of a roof over the engine and the driver's platform. Less obvious was the use of eight dry cells for the make-and-break ignition instead of the earlier wet cells.

There is some evidence that the 17–30 was a commercial success, although it was not, as has often been claimed, the first tractor to be produced on a commercial scale.

Meanwhile the 18–30 had been sold to a customer who kept it for 17 years. Later the tractor was bought back by the company, restored and presented to the Smithsonian Institution, where it is displayed in the Farm Machinery Section.

FORD EXPERIMENTAL TRACTORS

Henry Ford's spectacular success in the tractor industry began in 1917, when the Fordson arrived to dominate the market, but his ideas about farm mechanisation had started much earlier.

He was born in 1863, the son of an immigrant from Ireland who had left because of the potato famine to settle on a small farm in Dearborn, Michigan. Life on the family farm must have been an important influence on Henry Ford's ideas for developing a tractor which would ease some of the burden of work on the land.

'I have walked many a weary mile behind a plow and I know all the drudgery of it', he said in his autobiography in 1922. 'What a waste it is for a human being to spend hours and days behind a slowly moving team of horses.'

His interest in machinery, and his lack of enthusiasm for the farm, took Henry Ford to Detroit in 1879. He had several jobs there, while he gained experience in industry and acquired a reputation as a talented self-taught engineer. His work brought him into contact with internal combustion engines, which were still a novelty, with European technology firmly in the lead. He began to experiment in his spare time and, by 1894, had built at least one engine which worked reasonably well. This was followed in 1896 by Henry Ford's first car, which was belt-driven and equipped with two forward speeds but, at least on its first outing, no reverse gear and no brakes!

By 1905, after some earlier setbacks, Henry Ford was at the start of what was probably the most remarkable success story in industrial history. The Ford Motor Company was established in a new factory on Piquette Avenue, Detroit, and was beginning to make substantial profits. Ford cars were gaining a reputation for good design and competitive prices in a market which was expanding rapidly.

In spite of the demands made by his growing company, Henry Ford was already putting some effort into his ideas for power farming during 1905, although details are not available. The first experimental tractor was completed in 1907, designed and built to Ford's ideas mainly by Joseph Galamb, one of the most talented and trusted of Ford's engineering team. Galamb's involvement in the tractor project indicates the importance attached to it. He was born in Hungary and had worked in Germany before joining Ford in 1905. Later Joseph Galamb was closely involved in the development of the Model T car and the Fordson tractor project.

The 1907 tractor looked crude, but it included some important indications of Henry Ford's ideas of tractor design. An obvious feature of the tractor was its small size and light weight. Most American designers at that time were building heavyweights which were too large and too expensive for the smaller farms. The first Ford tractor weighed only 1500 lb, with a transverse engine-mounting which helped to keep down the overall length.

Another feature of the first Ford tractor was the way existing, readily available components were used in its construction. The engine was a four-cylinder, 24-hp unit built for the Model B car. The Model B was manufactured between 1904 and 1906 and was designed for the top end of the market.

Henry Ford at the wheel of what was probably the first experimental tractor built by his company in 1907.

Left and above. *More of Henry Ford's experiments in tractor design.*

The same engine also powered a racing car, which set a new American speed record in 1904, covering 1 mile in 39.5 seconds.

When the Model B car went out of production in 1906 it was replaced by the K. This was also an expensive car and it contributed radiator, steering gear and front wheels to the prototype tractor. The rear wheels of the tractor were from a binder.

Presumably Henry Ford was not satisfied that the 1907 design was suitable for the market, as it was never put into production. In any case his resources were becoming heavily involved in the development of a new car, which was to become the legendary Model T, the 'Tin Lizzie'.

More experimental tractors were built at intervals, probably 50 or more before the Fordson project was started in 1915. The experimental tractors appeared to vary considerably in design while following consistently the Ford theme of lightness and small size essential for a design cheap enough to bring tractor power to the smaller farm.

Henry Ford also continued to base his experimental tractors on existing car components wherever possible, so that parts of various Ford models are identifiable, including Model T engines and radiators.

SAUNDERSON FOUR-WHEEL TRACTOR

One of the outstanding pioneers of the early development of farm tractors in Great Britain was Herbert Saunderson. His company at Elstow, Bedfordshire, grew to become the biggest manufacturer and exporter of tractors outside the USA.

H.P. Saunderson was born at Cardington, Bedfordshire, the son of a baker. After serving an apprenticeship as a blacksmith, he went to Canada where he developed an interest in farm machinery. Before returning from Canada in 1890, he established a contact with either the Massey or the Harris Company. Exactly which company is not clear, but it is of little importance as the two merged to become Massey-Harris in 1891.

Back in England, Saunderson set himself up as an agent for Massey-Harris. He had premises at Kempston, Bedfordshire, where he sold and serviced Massey-Harris machinery, especially their very successful binders. Saunderson also branched out into other equipment, including windmills and pumps, and, in about 1900, he built his first self-propelled vehicle, which was described as a small waggon or dog cart.

A two-wheel tractor followed, which was intended to be a fore-carriage for a horse-drawn implement such as a binder. This was later converted to a three-wheel unit. The Ivel three-wheel tractor was creating considerable interest in the early 1900s and probably helped to set a trend which Saunderson may have followed. He produced a succession of three-wheel tractors, some with three-wheel drive and with a load-carrying platform to double as a farm truck.

The tractor illustrated was one of the first four-wheel tractors built by Saunderson. This was in about 1908 and, within 3 years, the company had abandoned the three-wheel design completely. The engine for the four-wheel tractor was a vertical four-cylinder unit with automatic inlet valves. It was designed by Herbert Saunderson and, like all Saunderson engines from 1906, it was built to run on paraffin after starting on petrol.

At the front of the tractor much of the space was taken up by the cooling system. The massive radiator, made of a large number of vertical ½-in copper

One of the first four-wheel tractors built by Herbert Saunderson, with company demonstrator, Sidney Dickinson, at the wheel.

tubes, was a design used on other Saunderson tractor models at that time. Later it was made more compact and was covered by a cowling.

At the rear of the tractor was a flat area where a load-carrying platform could be fitted. This idea, like the three-wheel tractor, was favoured by Saunderson for several years but abandoned soon after 1908.

The four-wheel design was developed at a time of great activity for Saunderson. He had recently started work on his own aeroplane and was building a range of ploughs and other farm machinery, boats with weed-cutting machinery for clearing rivers and canals, and also windmills to pump water and to power barn machinery. In addition, the company was exporting tractors as far afield as the USSR, South America and Australia, and adding more new models to the range so that, in 1910, the company catalogue listed seven different tractors.

RUMELY MODEL E OILPULL

By any standards the Model E was a great tractor – great in reputation and great in size and power. It is one of the best examples of the big tractors built in the USA to challenge the supremacy of steam on the big acreages in the western states and Canada.

Rumely tractors were built at La Porte, Indiana, where Meinrad Rumely had a blacksmith's business in the 1850s. Rumely was born in Germany, but had emigrated to the USA with other members of his family. The blacksmith's shop expanded to become a factory, building farm equipment and agricultural steam engines.

In 1908, the company made their first moves towards the fast-growing tractor market when John Secor joined Rumely to work on a new oil engine. Secor had a great deal of experience in engine design, including the development of water-injection systems. His influence on the Rumely tractor engine lasted almost 20 years.

The Model E went into production in 1911, soon after the 25–45 Model B, which had been the first Rumely tractor on the market. The E was rated as a 30–60 tractor, i.e. it should give 30 hp at the drawbar and 60 hp on the belt, and tests showed that it could achieve these outputs with power to spare. When the E was tested in the 1911 Winnipeg Agricultural Motor Competition, it produced 67.9 hp in the brake test. Nine years later in Nebraska Test No. 8, the maximum output on the belt was 75.6 hp and the highest drawbar figure almost 50 hp.

To achieve this performance, Rumely used a twin-cylinder horizontal engine with a 10-in bore and 12-in stroke. In the Nebraska Test, the engine produced its maximum output at a leisurely 378 rpm, one of the lowest engine speeds recorded in the test series. The cost of producing the maximum power was a fuel consumption averaging up to 10.7 gallons per hour of paraffin, which was also one of the highest figures recorded at Nebraska.

Another outstanding figure in the test report on the 30–60 was 5.2 per cent wheelslip in the drawbar assessment. This highly efficient performance was no doubt helped by the 26 000-lb weight of the tractor – there was no attempt to measure soil compaction in the tests.

Features of the engine design included a special carburettor with water injection, make-and-break ignition and induced draught cooling through the massive rectangular tower at the front of the tractor. A single forward gear ratio permitted travel speeds up to a maximum of a little less than 2 mph.

Rumely kept the 30–60 OilPull in production until about 1923. During this time, the company concentrated on powerful tractors which were heavy and rather dated in design, and an attempt to enter the lightweight tractor market in 1916 made little impact.

During the commercial life of the 30–60, the name of the company was changed to the Advance-Rumely Thresher Company, as the original company failed and was reorganised.

The mighty Rumely 30–60 from the Western Development Museum collection in Canada.

TWIN CITY 60-90

For the wealthy prairie farmer who wished to impress his neighbours, the Twin City 60–90 offered considerable possibilities. It also offered the capacity to cope with a big acreage.

Twin City tractors were built by the Minneapolis Steel and Machinery Company of Minneapolis, Minnesota. The company had been formed in 1902, following several mergers of smaller concerns, one of which was the Twin City Iron Works, which gave its name to the tractors.

Tractor production on a commercial scale appears to have started with contracts to build engines and complete tractors for other manufacturers. The first tractor to be marketed under the Twin City name was probably the 40, which was available in 1910. This model provided the design pattern for a range of Twin City tractors which were in production for about 12 years, with exposed engine, cylindrical radiator, a canopy and heavyweight proportions.

The 40 tractor had a four-cylinder engine with $7\frac{1}{4}$-in bore and 9-in stroke. The same cylinder dimensions were used for the engine in the 60–90, but there were six of them to develop the 90 hp which the manufacturer guaranteed.

When the 60–90 was being designed there were few signs of economy in its dimensions. The weight ex-works was 28 000 lb, more than 12 tons, and in working trim this would have been increased considerably with the fuel tank and radiator each filled with about 100 gallons.

While the 60–90 offered plenty of weight and power, there was little attempt at sophisticated engineering in its design. The drive to the 7-ft-diameter rear wheels consisted of a single-speed gearbox with a spur gear to the live rear axle. The maximum forward speed was 2 mph with the engine running at a leisurely 500 rpm.

The 60–90 was not the biggest tractor on the market, but it was the biggest built by the Minneapolis Steel and Machinery Company. It remained in production until the early 1920s, when changing economics brought the age of the heavyweights to an end until very big tractors became fashionable in the 1970s.

One of the biggest of the prairie heavyweights, the Twin City 60–90.

FAIRBANKS-MORSE 30-60

Fairbanks, Morse & Company was one of the large number of engineering concerns started during the period of massive industrial expansion in the nineteenth-century USA, which made an appearance later as a tractor manufacturer.

The company's interest in farm tractors was a logical development from the engine-manufacturing business which had been built up during the 1890s. Fairbanks-Morse engines were originally based on Charter gas-engine design and the range expanded to include small, single-cylinder horizontal engines which were among the market leaders. Later, the company manufactured very large diesel engines for powering generators and also marine diesels of up to 13 000 hp. Fairbanks, Morse & Company is still in business as a division of the Colt Industries organisation.

Experiments with tractor design started in about 1903, with a prototype based on a Morton chassis with a single-cylinder Fairbanks engine. There was apparently no commercial development from this project and the company did not begin marketing tractors until 1910, when the Fairbanks-Morse 15–25 was first shown in Canada.

Most of the tractors built by Fairbanks-Morse were designed for the big prairie farms, where pulling power was an important sales feature. The 30–60 model was certainly of this type, with 28 000 lb of weight to keep the 6 ft-6 in driving wheels firmly on the ground.

The power unit for the 30–60 was a twin-cylinder horizontal engine with $10\frac{1}{2}$-in cylinder bore and 12-in stroke. The two cylinders developed their power at 350 rpm, with a cooling system which consisted of a typical front-mounted tower, with 200-gallon capacity, through which air movement was induced by the engine exhaust.

Tractors of the 30–60 type represented a brief, but important and impressive, chapter in the history of farm mechanisation in the USA and Canada. The 30–60 went out of production in 1914, after only a small number had been built, and the company put its efforts into smaller tractors, some of which were sold under the Fair-Mor name. Fairbanks-Morse withdrew from the tractor market in about 1918 to concentrate on engine development and production.

This Fairbanks-Morse tractor is in the Ontario Agricultural Museum collection.

ALLIS-CHALMERS 10-18

The Allis-Chalmers organisation started in 1901 as a result of a merger of four companies in the heavy engineering industry. The first part of the new company's name was contributed by the E.P. Allis Company, a Milwaukee engineering concern which began as a manufacturer of equipment for the Wisconsin flour-milling industry in 1847. A Chicago company, Fraser and Chalmers, provided the remainder of the Allis-Chalmers name.

The new company continued to concentrate on heavy engineering and expanded successfully into steam turbines and generating plant. Then, in 1913, General Otto Falk became president of Allis-Chalmers and brought an important change of policy. He decided to spread the group's interests more widely and he chose the tractor industry as one way to achieve the diversification he wanted.

General Falk had a personal interest in farming and this probably influenced his decision to take his company into the tractor business at exactly the right time. The American tractor industry, in 1913, was approaching a period of massive expansion as the War forced the pace of mechanisation on the farms.

Although the General's timing was right, his company's first tractor was not a success. This was the 10–18, which started small-scale production in November 1914 and came on to the market early the following year, to be advertised as 'the only tractor that has a one-piece steel heat-treated frame –no rivets to work loose – will not sag under the heaviest strains'.

Allis-Chalmers chose a three-wheel design for their first tractor, with the unusual arrangement of having the single front wheel in line with the right-hand rear wheel. Although this looked odd, it had the advantage of keeping the front wheel in the furrow bottom while ploughing.

A horizontally-opposed twin-cylinder engine was used as the power unit for the 10–18. This was started on petrol, but switched over to paraffin when the running temperature had been reached. The main fuel tank was fixed against the inside of the mudguard, close to the driver's left knee.

The transmission was simple and basic. The gearbox provided only one ratio forward and one reverse, and the final drive was by pinions fitted with small rollers, which engaged in the teeth of the large ring gear on the inner circumference of each driving wheel. Although this arrangement of exposed drive to the wheels was used by many tractor companies at that time, it had considerable disadvantages. The ring gears were exposed to mud or dust, according to the ground conditions, and were also liable to collect stones, which could cause serious breakdowns in the field.

Demand for the 10–18 remained small and other manufacturers with more advanced designs gained most of the benefit of the wartime tractor

sales boom. In spite of this, Allis-Chalmers stayed firmly in the market and, in 1919, invested heavily in a new foundry and assembly building for the 10–18 and a smaller model, the 6–12, which was launched in 1918.

The 10–18 tractor remained in production, apparently, until 1920 and its replacement, the 12–20, came on to the market in 1921.

Allis-Chalmers 10–18 three-wheel tractor.

AVERY 25-50

There were two tractor companies in the USA with the name 'Avery'. One of these was the B.F. Avery Company of Louisville, Kentucky. The other was the Avery Company of Peoria, Illinois. There was apparently no connection between the two companies and this is just one example of the many instances of American tractor manufacturers having either the same company name or names sufficiently similar to cause confusion.

The Peoria company was at one time the larger of the two Avery concerns. It began in a small way in the 1870s making corn planters, originally at Galesburg, Illinois, and later at Peoria, where the company expanded into other types of farm equipment and, in 1891, began building a successful line of steam traction engines.

Steam engines brought the Avery concern its first major success, including an undermounted design which was an interesting break from traditional ideas. Avery also built a successful range of Yellow Fellow separators.

When Avery decided to move into the tractor market, their first attempts were not completely successful. In 1909, the company announced a tractor with a truck body, intended for road transport jobs as well as working on the land. This was an ingenious idea which possibly deserved more success than it achieved. In the following year, a more conventional tractor was entered in the 1910 Winnipeg Trials. The power unit for this new model was a single-cylinder engine, with 12-in bore and 18-in stroke, which developed just over 20 hp on the brake test. This tractor failed to complete the trial and seems to have disappeared from the market soon afterwards.

Early disappointments failed to discourage the Avery company, which at that time was controlled by John B. Bartholomew, one of the most dynamic personalities in the industry. Success came in 1911, with a 20–35, followed by a 12–25 in 1912 and a huge 40–80 in 1913. The 40–80 weighed 22 000 lb.

Two new models were announced in 1914, a small 8–16 and the highly successful 25–50.

Avery used twin-cylinder horizontally-opposed engines in their smaller tractors. Engines for the larger models were four-cylinder units which were effectively two of the smaller engines linked together. This system helped to keep some rationalisation of components at a time when new models were being introduced by the company with remarkable frequency. The engine for the 40–80 was two 20–35 tractor engines and, for the 25–50, Avery used four cylinders with the same dimensions as the two cylinders from a 12–25 tractor.

The four cylinders of the 25–50 developed their rated power at 700 rpm, with a maximum of 56 hp when the tractor was later tested at Nebraska. The cooling system for the 25–50 – like almost every tractor made by Avery until the early 1920s – was the round tower with draught induction.

Another distinctive Avery feature used on the 25–50 was the method of changing gear ratios. The gear-change mechanism consisted of sliding the engine forwards or backwards on a frame to select the forward or reverse ratio required.

An Avery 25–50 from the Western Development Museum collection.

INTERNATIONAL HARVESTER TITAN 10-20

International Harvester used the name 'Titan' for the range of tractor models built at their factory at Milwaukee, Wisconsin. Mogul tractors were built at the IH factory at Chicago.

The 10–20 was the smallest and also the most popular model in the Titan range. It weighed 5225 lb, while some of the bigger tractors from the same factory weighed four times as much.

Production of the 10–20 started in 1914 and continued through an important period of expansion and development in the world tractor market. During this 10-year period, the simple, rugged design of the 10–20 earned an exceptional reputation for reliability and also became extremely dated as design improvements were made on other tractors.

International used a twin-cylinder, horizontal engine for the 10–20. The bore was $6\frac{1}{2}$ in with an 8-in stroke and the engine developed its rated 20 hp at a leisurely 575 rpm. The engine operated on paraffin, with water added through a special metering device to prevent pre-ignition when working under full load.

Cooling-water for the engine was carried in the large cylindrical tank mounted over the front wheels. This held 39 gallons, which were circulated by a thermosiphon effect, boosted by steam pressure when the engine was hot.

Outdated features, such as a heavy steel-girder framework, a gearbox with only two forward ratios and an exposed chain and sprocket transmission, failed to stop the Titan's remarkable popularity. This applied to both sides of the Atlantic as the Titan was one of the most widely used of the many imported makes and models used in Great Britain during and just after World War 1.

One of the customers for the 10–20 Titan in Great Britain was the War Agricultural Committee for Kent, an organisation formed to supply manpower and equipment to help maintain food production in the county at a time when large numbers of farm workers, with their horses, had left the farms to join the Army. The report prepared by this Committee showed that 15 tractor makes had been tested during the War on farms in the county and, when the War ended, the fleet of 180 tractors operated by the WAC included 112 Titans.

International Harvester Titan 10–20.

WEEKS-DUNGEY NEW SIMPLEX

The original idea for the Weeks-Dungey tractor was developed by a farmer in the Weald of Kent, an area in Great Britain well known for orchards and hop gardens. The farmer, Mr Dungey, wanted a tractor compact enough to meet the requirements of hop-and fruit-producers. He took his ideas to William Weeks and Son, an agricultural engineering company at Maidstone, Kent, and they agreed to put a tractor into small-scale production.

The first tractors were produced in about 1915. The most prominent feature of the early design was a large cylindrical cooling tank, placed beside the driver, where it presumably provided warmth in cold weather, and over the rear wheels, where it provided weight for improved grip. The tank was linked to the four-cylinder engine by two pipes which circulated cooling water by convection.

Although production of the first Weeks-Dungey model was on a small scale, the results were sufficiently encouraging for improved versions to be developed. These resulted in the New Simplex. Details of the new model were announced in 1918 and it remained in production until 1925.

Changes included more protection for the pinion-and-ring drive to the rear wheels and a radiator to cool the engine. In the 1918 description, the weight of the New Simplex was quoted as 35 cwt and the dimensions were 8 ft long and 4 ft wide. The engine was a four-cylinder petrol/paraffin unit developing 22.5 hp and driving through a single-plate clutch and a gearbox with three forward ratios and a reverse. Although Weeks used several different engines, the American Waukesha appears to have been fitted to most of the New Simplex tractors.

One of the sales points of the New Simplex was a differential clutch which could be engaged by means of a lever. The advice to drivers was to

The original version of the tractor built by Weeks to the design of Mr Dungey.

Weeks-Dungey New Simplex.

engage the clutch when extra drawbar pull was required, as this produced a rigid axle. The clutch had to be disengaged when turning.

Weeks produced costings for the new model, based on the ploughing performance achieved in 'Kentish soil', They claimed that the tractor would pull a three-furrow plough with 10-in furrows working at 2.5 mph, and thus cover 15 acres in 6 days. Fuel consumption was recorded as 3 to 4 gallons an acre, plus 0.3 gallons an acre of lubricating oil.

Costs for the 6 days were:

Wages for driver and ploughman	£3 0s 0d	(£3.00)
60 gallons of paraffin	£4 10s 0d	(£4.50)
3 gallons of petrol	£0 9s 9d	(£0.49)
5 gallons of lubricating oil	£1 0s 0d	(£1.00)
Total	£8 19s 9d	(£8.99)
Cost per acre	£0 12s 0d	(£0.60)

WATERLOO BOY AND OVERTIME MODEL N

The Waterloo Boy Model N was one of the most important tractors ever produced. Although its design was old-fashioned and the sales performance modest, the Waterloo Boy achieved a more prominent place in tractor history than most of its rivals.

The Model N arrived on the market in 1916 and, when production ended 8 years later, less than 20 000 had been manufactured. Some of these were exported to Great Britain, where the tractor was sold as the Overtime Model N.

When the Waterloo Gasoline Engine Company of Waterloo, Iowa, began producing the Model N, its design was already becoming dated. It was based on an earlier Waterloo tractor of about 1913 and consisted of a massive frame which carried the fuel tank, radiator, engine and driver's platform as individual units.

One of the more up-to-date features of the Model N was the use of Hyatt roller bearings in place of the plain bearings of earlier Waterloo tractors. There were 12 roller bearings in all, on the cooling fan, rear axle, countershaft and transmission gears. The engine, with no cover to keep out the worst of the weather, was a twin-cylinder, horizontal unit, starting on petrol and running on paraffin. The transmission included two forward gear ratios and a reverse, a hand-operated cone clutch and final drive by open spur gear and pinion. The engine developed 25 hp on the pulley at 750 rpm and a drawbar pull of 2900 lb.

Waterloo Boy tractors, sold in Great Britain under the Overtime name, earned a reputation for sturdy reliability. They were used in the British campaign to increase home-grown food production during World War 1 and were also the beginning of a British development which eventually influenced the design of most of the world's tractors.

The Belfast agent for Overtime was a dynamic young business-man with a reputation as a talented engineer. His name was Harry Ferguson and the Overtime brought him his first direct contact with tractors.

Experience with the Overtime, and with other

An early Waterloo Boy tractor.

tractors he worked with, persuaded Ferguson that there must be a better way to attach implements to tractors. He began the development work which eventually produced the Ferguson System of hydraulic three-point linkage with automatic draft control, now standard equipment on most modern tractors.

The Model N also played an important part in bringing the John Deere company into the tractor business. John Deere, the man who founded the company, was born in 1804 in Vermont and trained as a blacksmith. He moved to Grand Detour, Illinois, in 1836 and, in the following year, built his first plough. His plough-making business expanded rapidly to meet the demand from settlers who were establishing farms in Illinois and further west. The fast-growing company moved out of Grand Detour to new premises in Moline in 1847 and continued to expand their production of ploughs and their growing range of farm implements.

The company stayed out of the steam engine business, but the expanding market for farm tractors proved to be more attractive. Deere & Company began experimenting with tractors in about 1912, building several prototypes, plus small numbers which were sold, but little progress was made until 1918, when Deere & Company bought the Waterloo Gasoline Engine Company for $2.1 million and became a significant force in the American

Rear view of a Waterloo Boy tractor showing the offset driving position.

tractor market.

The Waterloo Company had two Waterloo Boy models in production when the Deere takeover was announced. These were the Model N and a similar tractor with a reduced specification, known as the Model R. It was the N which was kept in production and held John Deere's place in the market until the famous Model D was announced in 1923.

In 1920, the Waterloo Boy Model N earned further space in the history books as the first tractor to complete the Nebraska Tractor Test. The tests were the result of the Nebraska Tractor Bill which was introduced in the State Legislature by Wilmot F. Crozier. The Bill was designed to ensure that tractors sold in the state were backed by adequate service facilities and that they met the manufacturers' claims for performance.

A test programme was devised by the Agricultural Engineering Department of the State University and the first test started early in the winter of 1919 on a Twin City tractor. This test was abandoned when snow made conditions impossible and new tests started again on March 31st 1920. This time the tractor in the test was a Waterloo Boy, which completed its programme on April 9th, the first of 64 tests carried out in that year.

SAUNDERSON UNIVERSAL MODEL G

Saunderson used the name 'Universal' for several quite different tractor models, which makes the company's history somewhat confusing. The confusion is increased because the name of the company changed several times, to Saunderson and Gifkins and also to Saunderson and Mills, as H.P. Saunderson brought in partners with finance to help the company survive occasional money problems.

Easily the most familiar Saunderson tractor model was the Universal which was in production from 1916, when the company was known as Saunderson & Mills Ltd. This model was based on a previous Universal, which had been designed with a forward driving position and a tall water tank and cooling system at the rear. The new model, which also included some gearbox modifications and other mechanical differences, allowed the driver much better rear visibility for ploughing.

The new version of the Universal was the first British-made tractor to meet a genuinely strong demand on its own market. The British tractor industry, which had done so much to pioneer new ideas, had previously been forced to concentrate on overseas sales for survival and had become the world's largest exporter of tractors in the process. When World War 1 started, Saunderson was the only British tractor company in a position to meet the rapid increase in demand. Other companies were operating on a very small scale and had either designed their tractors for the colonial market rather than British conditions, or dropped out of the tractor business in order to concentrate on Government contracts for military equipment.

The new version of the Saunderson arrived as people were beginning to realise that the War would last a long time and concern was increasing over the vulnerability of traditional sources of imported food. Suddenly, British farmers were being asked to raise output at a time when they had lost huge numbers of men and horses to the Army.

Tractor power was the obvious answer. King George V joined the move to mechanise and bought a Saunderson Universal in 1916. With it, he ordered a Saunderson four-furrow plough, a trailer and a little fuel-and-water cart. The tractor was delivered to the Royal Estate at Sandringham, Norfolk, by road. The journey of 80 miles from Bedford took 2 days and presumably ensured that the engine was well run in by the time the tractor was handed over to its new owner.

Saundersons benefited from the publicity the Royal order received. The King and Queen were

Saunderson Universal moving timber.

taken out to one of the Estate farms to see the tractor at work. According to a contemporary report, the Queen asked the ploughman if he liked the new tractor and was told it was a great improvement because, with a horse, he had to walk all day, but now he could sit on a seat. According to one published report, the tractor and plough at Sandringham did the work of ten horses and five men.

In 1917, the British Board of Agriculture placed an order for 400 Saunderson Universal G tractors, some – or possibly all of them – to be supplied with Saunderson four-furrow ploughs. This was a huge order for the company to cope with. To increase production sufficiently, much of the component manufacture was placed with outside subcontracted firms.

The Universal was a sturdy tractor with a good reputation for reliability and pulling power. The engine was a well-tried, twin-cylinder, vertical unit made by Saunderson and operating on paraffin. It was rated initially at 20 hp but was later described as being 25 hp.

H.P. Saunderson was a man who had made the most of many opportunities to sell his equipment throughout the world and to increase his business. He built his company up to become the biggest and most successful tractor manufacturer in Great Britain, and probably the largest in the world outside the USA. He also had a licence agreement with a French company to assemble the Universal for sale under the Scemia trade name. This agreement was apparently signed before World War 1, but there is little evidence that it became effective until the War ended.

The War and its aftermath brought further opportunities, but problems as well. One opportunity, which Saunderson used effectively, was the

Saunderson built this trailer for fuel and water for the Royal farm at Sandringham.

This was the post-War Saunderson design which was taken over by Crossley.

wartime publicity he earned by running a 240-acre heavy-land farm entirely by tractor power. At a time when most people in Great Britain believed that at least some horses were essential on a farm, H.P. Saunderson was able to supply data to the press on the performance of his equipment.

One of the post-War problems was the very competitive pricing policy which Henry Ford had achieved with the Fordson, which was making life difficult for everyone else in the tractor industry. The British market went into recession again after the War and there were difficulties in many of the export markets which Saunderson had once built up with such care and success.

Saunderson's Universal was given a rather meagre face-lift after the War, distinguished mainly by a styling change to the top of the radiator. At one time a 3-year warranty was offered.

The action taken was not enough to keep the company in business, when what was needed was a completely new, much more modern and competitive tractor.

Perhaps Mr Saunderson was losing enthusiasm. In 1924, he sold his business to the Crossley Company of Manchester, taking a year's consultancy as part of his payment. Crossley made some quite minor changes to the Universal G and tried to sell it as a transport tractor for road haulage work. This was not a success and brought tractor production at the Elstow Works to an end. H.P. Saunderson retired to an estate near Bedford, where he farmed with Universal tractors.

INTERNATIONAL HARVESTER 8-16 JUNIOR

One of the most popular tractors produced in the USA during World War 1 was the International Harvester 8–16. It was also one of the most distinctive in appearance, with a mid-mounted radiator which allowed a neatly sloping bonnet-line over the engine.

Large numbers of 8–16 tractors were shipped to Europe soon after the War ended, particularly to Great Britain where the tractor was known as the International Junior.

Production started in 1917 and continued until 1922. In some ways the design was already old-fashioned when the tractor was launched, with a completely exposed chain and sprocket drive to the rear wheels and a separate frame to which the components of the tractor were attached.

But the 8–16, or Junior, also had one feature which put International ahead of its competitors. This was a power take-off (p-t-o) shaft, which was available as an optional extra and could be used to drive trailed equipment. The p-t-o was not an International invention as it had been used on the British Scott tractor in 1904 and, a few years later, had appeared on tractors in France. The significance of the p-t-o on the 8–16 was that this was the first time the idea had achieved any significant commercial success and, in 1921, International made the p-t-o a standard fitting on their new 15–30 tractor.

International made a four-cylinder engine for the 8–16, which achieved 18.5 hp in the maximum load test at Nebraska and a drawbar pull of 1532 lb. The average engine speed in the tests was 1007 rpm. The weight of the 8–16 was approximately 1.5 tons and it had a top speed of 4 mph in third gear.

In the highly competitive tractor market of the early 1920s, International Harvester introduced a series of important new models, including the 15–30, 10–20 and, in 1924, the Farmall. These new models helped to put the 8–16 off the market after a production run of only 5 years.

International Junior with a pair of seed drills.

FORDSON MODELS F AND N

Henry Ford's Model F was the most important and successful tractor in the history of power farming. The Fordson F and N models brought the benefits of tractor power to more of the world's farmers than any other tractor before or since and it dominated the world tractor scene in a way which is unlikely to be repeated by any other manufacturer in the future.

Although Henry Ford did not personally design the Fordson, he was the man who made its success possible. It was Henry Ford who had the vision to foresee the possibilities for a simple, low-cost tractor manufactured in huge numbers and it was Ford who had the talented engineers, the mass-production know-how and the immense financial resources which could turn the vision into a reality.

Experimental design work, which had produced an assortment of prototype tractors since 1907, became more purposeful in 1915, when Henry Ford put more of his best engineering staff behind the tractor project. These included Joseph Galamb, who had been involved in the earlier development work, Eugene Farkas, the chief of design, and Charles Sorensen, who had become one of Henry Ford's most trusted executives in commercial matters.

At about this time, a new company, called Henry Ford & Son, was formed as a way to keep the tractor project separate from the Ford Motor Company. All the shares in Henry Ford & Son were held by members of the Ford family, whereas there were several outside shareholders with substantial interests in the main company.

Development work progressed, with Eugene Farkas contributing the idea for building the engine, gearbox and rear axle as stressed castings which could be joined together to form a self-supporting main unit. This meant that the Fordson could be built without a separate frame, saving cost and simplifying the manufacturing process. This was an improvement on the frameless design which had already been developed for the Wallis tractor.

An early Fordson Model F at the Science Museum, London.

Henry Ford authorised the production of a batch of 50 to 60 tractors based on the Farkas design to be used for field test work. Two of these were shipped to England at the request of Percival Perry, the chief executive of the Ford Motor Company in Great Britain. The British Government was anxious to increase the number of tractors available to help boost food production from British farms. The two Fordsons performed well in the tests and the fact that they were designed to suit mass production and would be economical to operate appealed to the British authorities.

Charles Sorensen travelled to England in the spring of 1917 to make arrangements to put the Fordson into production under a Government-backed scheme. This project was abandoned whilst still in the planning stages, following an official switch of priorities. Instead it was agreed that Henry Ford would make immediate arrangements to put the tractors into production in a factory at Brady Street and Michigan Avenue, Dearborn, against a firm order from the British Government. Production arrangements were completed with a remarkable speed which the Ford resources made possible; the first of the new tractors was completed on October 8th 1917 and the initial batch of 7000 ordered by the British Government was completed within 6 months.

The Model F tractors, which were shipped to Great Britain in 1917 and during the first months of 1918, all had four rectangular slots at each side of the radiator. This is a distinctive feature of the first production batch, as the side panels were plain from 1918 until 1932, when the Fordson name was cast on the panels in raised lettering.

Production of Model F tractors increased rapidly to a peak of more than 100 000 in 1923 and again in 1925. Almost 750 000 Fordsons were built between 1917 and 1928, the biggest production run achieved by any tractor model. During this period the Fordson at times accounted for 75 per cent of all tractor production in the USA and held more than 50 per cent of the world tractor market.

The biggest single customer for Fordson tractors at that time was the Russian Government, with a huge programme of agricultural reconstruction to cope with after the Revolution. By 1926, more than

A Ford publicity picture for the Dagenham-built Fordson.

Tractor number 150 000 from the Dagenham production line.

25 000 Fordsons had been shipped to the USSR and, in the following year, it was claimed that 85 per cent of the Russian tractors and trucks had been built by Ford. Later Henry Ford co-operated with the Russian authorities in setting up a factory in the USSR to manufacture a tractor based on the Fordson.

In 1929, Fordson tractor production was closed down in the USA and transferred to Cork, in the Republic of Ireland, where some assembly and local manufacture had been in progress between 1919 and 1922. Changing factories provided an opportunity to make some changes to the tractor. The bore of the four-cylinder engine was increased to $4\frac{1}{8}$ in to raise the power output to about 25 hp instead of 19 hp. The updated tractor, known as the Model N, was equipped with a water pump and a more conventional magneto.

Henry Ford's choice of Cork as the location for a large tractor-manufacturing plant was probably based more on emotion than on economic logic. His father had sailed from Cork to start a new life in the USA after the Irish potato famine of 1846 and Henry Ford was probably pleased to be able to use some of his wealth to provide employment in the area. The problems facing the Cork operation were substantial. There was a lack of local skilled labour, almost all raw materials had to be imported and, apart from the handful of tractors supplied to the

Irish market, the entire output from the factory had to be exported.

After little more than 4 years of operation in the Republic of Ireland, the Ford tractor plant was transferred again, this time to England, where production started in 1933 at Dagenham, Essex. Tractor output at Dagenham rose to a peak, for the Fordson N, of more than 26 000 in 1942 and 1943, at which time Ford held more than 90 per cent of wartime tractor production in Great Britain.

Fordson production had started in 1917 to help increase food production from British farms in World War 1. In World War 2, when food production in Great Britain had again become a matter of great importance, the Fordson was easily the most important source of tractor power on British farms.

SAMSON MODEL M

The commercial battle between Ford and General Motors, the two largest international giants of the motor industry, has concentrated on cars and trucks. General Motors has had little direct involvement in the farm tractor industry, apart from a 5-year period from 1917.

In 1917, Henry Ford put his new tractor into production for the first time. In the same year, General Motors bought their first stake in the farm tractor business when they bought the Samson Sieve Grip Tractor Company of Stockton, California. They increased their challenge in the tractor industry when they bought additional manufacturing capacity at Janesville, Wisconsin, in 1918, and the rights to a small, four-wheel motor cultivator.

With the Sieve Grip Company, General Motors became the manufacturers of the Sieve Grip tractor, an unconventional three-wheel design named after openings in the wheel rims intended to improved pulling efficiency on loose soil. Although the Sieve Grip, originally introduced in 1914, was fairly successful, it was too expensive and too specialised to offer General Motors the big production volume they needed to compete with Ford. The Sieve Grip remained in production while General Motors developed a completely new design. The new tractor was the Samson Model M, announced in December 1918 and manufactured in the Janesville factory.

The price of the new Samson tractor was not quite as low as that of the Fordson, but the difference was largely offset by the better specification of the Samson compared with the basic, no-extras Fordson.

Both tractors used basically similar four-cylinder, water-cooled, petrol/paraffin engines. The Samson engine had a slightly bigger capacity, with 4-in by $5\frac{1}{2}$-in cylinders, compared to the 4-in by 5-in of the Fordson.

In 1920, the two tractors competed on level terms in the Nebraska Tractor Tests, with the Fordson taking Test No. 18 and the Samson taking No. 27. The test results suggest that the Samson outperformed its rival.

In the belt tests, where maximum performance was recorded for 1 hour, the Samson developed 19.39 hp and the Fordson returned 19.15 hp. In the same test, the Samson used nearly 10 per cent less fuel and only 2.62 gallons of water per hour instead of 3.35 gallons for the Fordson.

A significant difference between the two tractors was the weight. The Nebraska weighbridge recorded 3300 lb for the Samson and 2710 lb for the Fordson, and the extra pounds probably helped the Samson in the drawbar tests. During the 10-hour rated load test, the wheelslip from the Samson was 14.5 per cent. The Fordson wheelslip was measured as 23.8 per cent, the second highest figure recorded in the 65 Nebraska Tests during 1920. In the 10-hour test the Samson pulled 1252 lb at 2.94 mph, while its rival managed only 886 lb at 2.57 mph, with the Fordson using almost 10 per cent more fuel per hour.

On this evidence, the Samson was a good tractor, but in commercial terms it was a failure against the Fordson. General Motors was losing money on the Samson and could not match Henry Ford's policy of aggressive price cuts to maintain sales volume.

By 1923, the battle was over. General Motors had pulled out of the farm tractor market and the Janesville factory was being converted into a Chevrolet assembly plant.

In the same year, Henry Ford's factories produced more than 100 000 tractors, more than 50 per cent of world production.

The Samson Model M, built by General Motors to rival the Fordson.

The Samson tractor illustrated, one of very few still in existence, is preserved – perhaps appropriately – in the Greenfield Village and Henry Ford Museum at Dearborn.

MASSEY-HARRIS 12-25

The 12–25 was one of the early attempts by the Massey-Harris Company to break into the farm tractor market. The company had grown rapidly and successfully in the farm machinery business, with a particularly strong position in harvesting equipment, but without a tractor to sell.

It became increasingly evident that a tractor would be a valuable addition to the Massey-Harris product line. Tractor sales had increased rapidly after about 1910 in the USA and Canada and Massey-Harris had failed to take advantage of the expanding market.

In 1917, the Canadian company made its first move into the tractor business through an agreement with the Bull Tractor Company of Minneapolis, Minnesota. Under the terms of the deal, Massey-Harris imported the three-wheel Big Bull tractor into Canada to be sold as part of the Massey-Harris range of equipment.

The Big Bull proved to be a commercial failure for Massey-Harris, partly because of design problems and partly because the Bull company was facing production problems. In 1918, Massey-Harris made a second attempt at the tractor market, this time through a deal with the Parrett Tractor Company of Chicago.

During the wartime years of booming sales, tractors designed by brothers Dent and Henry Parrett achieved considerable success. Most Parrett tractors featured extra-large diameter front wheels, claimed to need less power when running over uneven ground and to give less compaction on soft soil. The big diameter was also said to reduce bearing wear, because the rotation speed was slower and because the hubs were higher above the mud.

The 12–25 with 46-in-diameter front wheels was probably the biggest success in the Parrett range and was one of three models involved in the deal with Massey-Harris. The same model was also exported in small numbers to Great Britain where some some were sold under the Parrett name and others were marketed under an arrangement with John Wallace of Glasgow and were known as Clydesdale tractors.

Engines for the Parrett and Massey-Harris versions of the 12–25 were supplied by Buda. These were four-cylinder power units, operating at up to 1000 rpm. This was exactly the right kind of engine according to the Massey-Harris brochure on the 12–25.

Massey-Harris 12–25.

'Careful tests have convinced us that a 4-cylinder engine running up to 1000 revolutions per minute, is the ideal engine for tractor use. Slow speed and fewer cylinders mean a heavy engine, unsteady power, excessive vibration, and short life', the brochure explained.

'A speed in excess of 1000 would mean increased wear and vibration, while more cylinders mean added complication without corresponding increase of efficiency.'

The engine was water-cooled with the distinctive side-facing radiator. The tractor weighed 5200 lb and the overall length was 12 ft. There were two forward gear ratios and a reverse, with a maximum travel speed of 2.68 mph.

As the early 1920s brought more competitive and less profitable trading conditions, the Parrett company encountered financial problems and ceased operating in about 1922. Meanwhile, Massey-Harris were finding it difficult to build up the sales volume with what had become a dated design. The Parrett tractors failed to establish Massey-Harris firmly in the tractor industry and the project came to an end in about 1923 when production at the Weston, Ontario, factory was halted.

GLASGOW

As a rule farmers tend to be suspicious of unconventional ideas and equipment unless the advantages are obvious or can be easily proved. The suspicion increases when the novelty is much more expensive than the established alternatives.

The Glasgow tractor was unconventional and expensive and it was a commercial failure. The failure was to some extent unfortunate as the curious design had advantages in some situations and the Glasgow was an ambitious attempt to establish tractor production on a large scale in Scotland for the first time.

Glasgow tractor production started in 1919 on a 25-acre industrial site at Cardonald, which had recently been released by the Government from wartime production of ammunition for the British Army. The project was backed by a consortium of Scottish companies, which included the Wallace

In spite of some improvements made in the later version of the Glasgow, the project was never a commercial success.

An early version of the Glasgow three-wheel drive tractor.

Farm Implements Company of Glasgow, a leading British farm machinery manufacturer. Wallace apparently took a controlling interest subsequently.

Press announcements claimed that production was scheduled to rise to 5000 tractors a year. A sales contract was signed with the British Motor Trading Corporation to distribute Glasgow tractors in Great Britain and colonial territories.

The Glasgow was designed by W. Guthrie, who chose a three-wheel configuration with the single wheel at the rear. The three wheels were of equal size and all three were driven. Instead of a differential in the drive to the front wheels, there was a system of ratchets to compensate for unequal wheel speeds while the tractor turned.

A Waukesha engine, imported from the USA, was used to power the Glasgow. This was a conventional four-cylinder unit rated at 27 hp, driving through a cone clutch and a gearbox with two forward ratios and a reverse.

Guthrie designed the Glasgow primarily as a pulling tractor to work in difficult conditions. The three-wheel drive system was intended to give improved pulling efficiency and the absence of a differential helped to limit wheelslip in wet conditions. The driver was perched on a sprung seat which stuck out beyond the back of the tractor so that he could reach the controls of a trailed plough. This meant positioning the steering wheel and gear and brake levers some distance behind the rear axle.

In its first production version, the shape and styling of the fuel tank and radiator were rounded and closely resembled the appearance of a Fordson. During its brief period in production, a second version was introduced with more angular styling at the front.

Production fell short of the company's plans and the project was soon in difficulties. One problem was the failure of the distribution company due to financial pressures. But the basic difficulty was lack of sales. The unconventional design inevitably resulted in higher production costs and the 1919 list price was £450. Even when reduced to £375, the Glasgow was still difficult to sell because of competition from the imported Fordson, which cost about £120.

There were many who regretted the end of the Glasgow project in about 1924. On steep land or in wet conditions, where an ordinary tractor lost adhesion, the Glasgow design was at its most effective. There appear to have been plenty of testimonials from delighted customers, especially in Scotland, who praised its performance in conditions where conventional tractors struggled.

Four years after production ended there was an attempt to revive the project with the formation of a new company called Clyde Tractors Ltd. The Glasgow would be launched again, said the press announcement. More efficient production methods would allow a price reduction to £250. But the revival remained a Scottish dream which failed to reach the production stage.

RENAULT GP, H1 AND HO

After World War 1 ended in 1918, the international flood of new arrivals on the tractor market included three from leading car manufacturers in France: Citroën, Peugeot and Renault. All three companies announced crawler tractors, which were attracting considerable interest in France at that time. Citroën also produced a small wheeled tractor for vineyard work.

Citroën and Peugeot made only a temporary appearance in the tractor market and later concentrated on cars and commercial vehicles. Renault made a more determined entry into the tractor industry to become the largest French manufacturer.

During the war, Renault had developed a successful small assault tank for the French Army and

A Renault GP tractor moving timber.

this project had given the company valuable experience of tracklaying vehicles. The wartime tank contributed a substantial amount to the design of the peacetime tractor, which first appeared in 1919 as the GP.

Renault's GP tractor weighed approximately 7300 lb. The engine was a four-cylinder unit developing 30 hp on petrol, with a transmission consisting of a cone clutch and a gearbox with three forward ratios and a reverse.

Experience with the GP tractor led to some modifications, particularly to the crawler track. These were incorporated in a new version of the tractor which was known as the H1, and which was apparently on the market in about 1920. The unusual tiller steering of the first GP model was retained in the H1, but with the tiller redesigned – like the handlebars of a bicycle – so that it could be controlled by two hands.

In spite of the difficulties in the European tractor market in the 1920s, caused partly by the low price of American imports, the Renault was sufficiently successful to encourage the company to introduce an additional model. It had become evident that the market for crawler tractors would not grow as rapidly as the demand for wheeled models. The new tractor announced in 1922 by Renault was, in effect, a wheeled version of the H1 and was known as the H0.

The H0 was smaller and lighter than the crawler tractor, weighing 4700 lb, and with a four-cylinder petrol engine which developed 20 hp at 1600 rpm. An unusual feature of the H0 was the use of epicyclic reduction gears in the rear wheels.

Both the H1 crawler and the H0 wheeled tractor inherited the general layout and styling of the original GP, which had been one of the most interesting and distinctive-looking tractors built in Europe in the post-War period. All three versions were based on an immensely strong steel-girder frame, with a front-mounted pulley driven directly from the crankshaft. The stylish bonnet or hood covering the engine conformed to the design used on some of the first Renault cars and trucks, which had become almost a trademark for the company's products. Behind the engine was the radiator, sloping forwards to give a lower profile and better forward

Renault H1 photographed in 1920 at Villacoublay, France, during trials.

The caption to this publicity picture from Renault says 'Journey in the woods of Medou, 1923'.

Renault's first wheeled tractor, the HO.

visibility. Behind and above the radiator was the fuel tank – cylindrical at first, but shaped later to fit the space available.

A new Renault tractor, the PE, produced in 1927, brought the first complete break from the old GP styling with a more conventional appearance, which was probably cheaper to build but also less attractive.

FIAT 700, 702 AND 703

The Fiat organisation in Italy is one of the world's largest manufacturers of motor vehicles and it is also one of the oldest. Fiat history dates back to 1899 and the company quickly demonstrated its engineering ability with a series of racing successes which included the 1907 French Grand Prix.

During World War 1, Fiat, like most European engineering companies, was heavily involved in military contracts. When the War ended the company quickly returned to car production, helped by more successes on the race track, including another French Grand Prix victory. Fiat's rapid growth in the 1920s was encouraged by the development of some highly successful small cars and also by a protectionist policy by the Italian Government which made it difficult for foreign manufacturers to establish themselves in Italy.

While World War 1 was still being fought, engineers at Fiat were working on the design for a new tractor. The project had been given substantial encouragement by the Government which wanted to ensure that there would be Italian tractors available to help build up the country's food production when the War ended. The Government also provided guidelines for the size and type of tractor they considered most suitable for Italy's main arable areas.

Fiat completed the field tests of their first tractor in 1918 and production began at the Corso Dante factory in Turin early in 1919. The new tractor was the 702, later modified to become the 703.

Although the 702 was the first tractor to be manufactured in volume in Italy, it was an outstanding success. Some of the 702s were exported to Great Britain where they performed well in direct competition with British and American models.

Two of the new tractors arrived in Great Britain in time for the National Tractor Trials which were held at South Carlton, Lincolnshire, in 1919. The tractors were accompanied by excellent sales literature, which included comprehensive information

Fiat 702.

on the performance the tractors were designed to give in a wide range of field operations and in varying conditions. For ploughing, the claimed output ranged from 1.25 acres an hour on light land with a six-furrow plough working at 8-in depth, to 0.225 of an acre an hour on hard land with one furrow set at 16-in depth.

Judges at the British trials praised the fuel economy and manouevrability of the Fiats and their performance in public helped to establish the 702 on the British market.

In Italy, the authoritative journal, *Corriere della Sera*, welcomed the new tractors as 'the highest expression of the efficient and profitable contribution Italian industry will make to agriculture in peacetime for the economical recovery and greatness of the Nation'.

The 702 was basically a conventional tractor, but with some unusual features. The engine was a four-cylinder, petrol/paraffin design, developing about 25 hp and driving through a gearbox giving three forward speeds and a reverse. Less conventional was the half-elliptical transverse spring over the front axle, the off-centre driving position for better visibility and the belt pulley mounted at the rear, close to the driver's seat. The pulley had three speeds, controlled by a hand-lever.

The 703 was similar to the original 702 in appearance and dimensions, but included some improvements to increase the performance at the drawbar. One of the changes was to fit a long drawbar attached near the front of the tractor underside. This meant that the effective hitch point for a draft implement was ahead of the centre of the tractor, which must have helped to improve wheelgrip in difficult conditions.

A second modification was the addition of a set of reduction gears in hubs of the driving wheels. These could be engaged in conditions where extra pulling power was required, but disengaged again – the job took five minutes according to Fiat – when the tractor was working normally.

According to the sales literature for the 703, the pulling power was increased from a maximum of 5292 lb at the drawbar for the original model, to 8268 lb with the new drawbar and reduction gears of the 703. A kit was offered to convert the early model to the improved specification because the tractor engine and other components were unchanged.

Of the 702 and 703 tractors, more than 2000 were built between 1919 and 1926. At this time, a new model, the 700, was announced. This was a more powerful tractor, but smaller and with a smaller, more efficient engine. The cylinder dimensions of the 25-hp engine of the 702 had been 105 mm × 180 mm, whereas the 700 engine had 90-mm × 140-mm cylinders developing 30 hp.

The 700 went into production early in 1927 and, with various modifications, remained on the market until 1942, with production totalling 4000. Some of the emphasis went out of the sales effort for the 700, partly because of economic and political problems affecting European trade in the 1930s and partly because Fiat was achieving substantial success with crawler tractors from 1932.

A Fiat 702 tractor modified for moving railway trucks.

Fiat 700 (left) and 703.

CLETRAC MODEL F

Rollin H. White was one of three brothers who established the family name in the American automobile industry in the early 1900s. Rollin was apparently most closely involved in design work and has been credited with much of the development of the White steam car.

In 1911, Rollin White designed a self-propelled disc cultivator which failed to get past the experimental stage. Later he and an assistant, Edward Ruck, worked on ideas for crawler tractor development, which resulted in a new steering mechanism. This was the controlled differential system which could be operated by a steering wheel instead of the usual steering levers.

A company was formed by Rollin White in 1916 to build crawler tractors with the controlled differential mechanism. The company was based at Cleveland, Ohio, and called the Cleveland Motor Plow Company. The name was changed in 1917 to the Cleveland Tractor Company. Tractors were sold under the Cleveland brand name until 1918 when the name 'Cletrac' was adopted.

Production started with the Model R, which was the tractor built to bring the controlled differential steering mechanism on to the market. This operated by slowing down the drive to one track which had the effect of speeding up the drive to the other track. The mechanism proved to work effectively in the field and became a feature of all Cleveland and Cletrac tractors. Later it was adopted by other manufacturers of crawler tractors and other tracked vehicles, including army tanks.

Cletrac
Cletrac
Cletrac

Top left. *A high-clearance version of the Cletrac Model F with controls extended to operate from the implement seat.*

Bottom left. *This 1921 photograph shows the narrow, standard and high-clearance versions of the Model F at the Cletrac factory.*

A modified version of the Model R tractor came on to the market as the Model H, to be followed in turn by the Model W. By accident or design, these first three model identification letters spelled R.H. White's initials.

From 1920 to 1922, the Model F Cletrac was in production. This was an attempt to build a lightweight, low-cost crawler tractor with a new track system designed to operate with minimum rolling resistance. The Model F was available with standard or narrow tracks, and in a high-clearance version for rowcrop work.

At that time Rollin White had made a brief return to the car industry and was manufacturing the Rollin car. The car and the tractor both used the same engine, which was made by the Cleveland company and was a four-cylinder, side-valve unit which provided 16 hp at the belt and 9 hp at the tractor drawbar and operated at 1600 rpm. Engines for the tractor were designed to operate on paraffin.

With an overall length of only 80 in and a weight of less than 2000 lb, the Model F was suitable for smaller acreages. The price in 1920 was $845, but this was later reduced. Apart from its small size, the most distinctive feature of the Model F was the appearance of the tracks. Each track was driven by a small sprocket positioned high enough towards the rear of the tractor to give a triangular outline.

Inside the circumference of each track was a chain or secondary track made up of a series of steel rollers. This had no driving function, but acted as a low-friction surface for the main tracks to run against. There were no lower track wheels, as the chain of rollers made these unnecessary.

The idea behind the ingenious track arrangement was to minimise working friction so that the tracks absorbed less power. The theory was sound, but dust and mud found its way into the roller chain during work and this led to a high wear rate.

During the short production life of the Model F, several companies produced front-mounted equipment to suit it. The tractor was also used with the controls extended to the rear, so that it could be operated from the seat of implements designed originally for working with horses. In spite of these indications of commercial success, Cleveland put their future efforts into larger tractors and did not attempt to re-introduce their roller chain tracks.

BRYAN LIGHT STEAM TRACTOR

When the internal combustion engine began making headway on the farm, not everyone welcomed its success. Farmers who bred working horses resented what they saw as a threat to their livelihood and the big companies in the steam traction engine business had a similar reaction.

Some of the steam-engine manufacturing firms, such as Case in the USA and Marshall in Great Britain, moved into the tractor business as soon as they were confident of establishing a long-term future there. Some companies simply went on building traditional steam traction engines, portables and road rollers until the market more or less disappeared.

There were also some manufacturers who realised that there were enormous advances in the technology of steam power which had never been used on the farm. These were the companies which transformed the traditional heavyweight steamer into a surprisingly effective competitor of the conventional farm tractor. Several companies experimented with steam tractors during the post-War decade, but only one manufacturer achieved a worthwhile commercial success.

The company which achieved some success was Bryan Harvester of Peru, Indiana. They took some

Bryan Light Steam Tractor.

of the ideas which had made steam cars a practical possibility in the early days of the motor industry and put them into a steam tractor which eliminated all the disadvantages which had put the traditional steamers on the road to the scrap heap.

A prototype Bryan steam tractor was apparently built in about 1920, but commercial production did not start until 1922. The Bryan company continued building steam tractors for about 5 years.

One objection to the old-fashioned steamer was the problem of maintaining a supply of bulky solid fuel, such as coal, logs or even straw. There was also the hard, dusty work of putting fuel on the fire and getting rid of the accumulating ashes. The Bryan tractor used paraffin, just like most ordinary tractors, with a vaporising burner into which the fuel was fed under pressure.

Much of the cumbersome bulk of the traction or portable engine was the large boiler where the steam was raised plus the supply of water to replace lost steam. A boiler explosion could be a disaster, occasionally causing loss of life. The boiler in a Bryan was a low-volume, high-pressure water tube, which was light, compact and very much safer than a traditional system. A condenser fitted to the tractor reduced water loss and the 60-gallon water tank carried on the tractor was claimed to be sufficient for a day's work.

An old-time steam-engine driver had to get up early in order to light the fire and raise steam pressure. A high pressure tubular boiler could be at working pressure from cold in a few minutes although, surprisingly, this was a sales feature which was not mentioned in the copy of the Bryan brochure which I have.

The brochure did stress the mechanical simplicity of the Bryan. The engine was a twin-cylinder unit with 4-in bore and 5-in stroke, operating at up to 220 rpm with Stephenson Link valve gear. A steam engine does not need a clutch or a complicated gearbox and there was no electrical system to cause problems. Bryan claimed there were just 39 moving parts in the tractor, only 13 of which were in the engine – and no fast-moving parts.

Bryan described their tractor as developing 20 hp on a steam rating, but it was probably equivalent to a 25- or 30-hp conventional tractor, with a 15-hp pull at the drawbar.

With so many positive features, it is surprising that the Bryan tractor failed to make more impact or to last longer. Perhaps it was simply too late, with steam already considered to be too old-fashioned to be acceptable. Perhaps there were technical problems in the field which the company failed to overcome.

There are still some optimists who believe there will be another steam age, in an energy-starved world where internal combustion has reached the end of its development. Presumably the Bryan, with its ability to operate on any liquid which will burn, suggests one of the ways a tractor of the future might operate.

HSCS

Nathaniel Clayton and Joseph Shuttleworth formed a highly productive and successful partnership in Victorian England. They were associated with much of the early development of steam engines for agriculture and their factory at Lincoln was one of the largest production units in the world for farm equipment.

Large numbers of the Clayton and Shuttleworth engines were exported and one of the most valuable markets for the company was Hungary. The customers in Hungary were the owners of the vast feudal estates which used large numbers of steam engines to power threshing machines. In order to develop the market, and to help service the equipment already in use, Clayton and Shuttleworth set up a small subsidiary company in Hungary.

In 1912, with political uncertainty in much of Europe and with markets for agricultural steam engines shrinking everywhere, Clayton and Shuttleworth decided to pull out of the Hungarian market. Their subsidiary company was taken over by Hofherr and Schrantz, a Hungarian partnership making farm machinery. The new company was called Hofherr, Schrantz, Clayton and Shuttleworth, or HSCS for short. The headquarters of the new company was initially at Kispest, but it was later moved to Budapest.

HSCS began building petrol engines in about 1919 and this development led to the design of the first HSCS tractor, which was completed in 1923. This was a dated design, with one of the company's single-cylinder petrol engines mounted on a mas-

This early publicity picture of the HSCS gives it a curiously foreshortened appearance.

sive steel frame and with exposed gearing.

The 1923 prototype was followed in 1924 by the model which brought HSCS into the tractor market for the first time. The new version was still basically old-fashioned, with a steel frame and a single-speed transmission; but the petrol engine used in the first HSCS tractor was now replaced by a single-cylinder semi-diesel engine.

Although the semi-diesel or hot-bulb engine was invented in Great Britain, very few British tractor manufacturers used it. It was also unpopular in the USA. There was much more interest in the semi-diesel in Germany and in Italy, and also in eastern Europe where HSCS was one of the first tractor companies to fit a semi-diesel in a production model. The advantages of this type of engine for agricultural use are the mechanical simplicity and reliability and the fact that a semi-diesel will burn almost any kind of liquid fuel, including waste engine-oil and low-grade fuels. The engines were usually single-cylinder, horizontal designs, with a slow operating speed.

The HSCS engine was rated at 14 hp and the tractor was sold for stationary work and for ploughing. The makers claimed that the tractor would plough a seemingly modest 4 acres a day to a depth of 8 in.

One of the 1924 tractors has been preserved at the Hungarian National Museum of Agriculture in Budapest.

HSCS retained the semi-diesel throughout a series of wheeled and crawler tractors, which earned a reputation for sturdy reliability rather than for advanced design. The company sent a tractor to the 1930 World Trials in Great Britain, but apparently with little commercial success.

When the communist régime took control in Hungary, the capitalist associations behind the initials HSCS were probably undesirable. In about 1951, the company name was changed to Red Star Tractor Factory and, from 1960, the products were sold under the Dutra brand name, which was derived from the words DUmper and TRActor. Dutra tractors were exported to several western countries as well as to other communist bloc countries. Some of the export tractors were equipped with Perkins engines from Great Britain, reviving the association with British technology which had brought the original company into existence.

More recently, the tractor plant has become part of the Hungarian Railway Wagon and Machine Factory. The products, at the time of writing, are the RABA-Steiger tractors built under licence from the American Steiger company, and these are providing the power required for the vast acreages of the State farms in modern Hungary.

1 This International Harvester Titan 10–20 was exported to Britain during World War 1.

2 (Above) *Saunderson Universal Model G.*

3 (Below) *International Harvester 8–16 Junior.*

4 (Above) *Fordson Model N built in 1933.*

5 (Below) *One of the green Fordsons built at Dagenham during World War 2.*

6 Fiat 702 built in 1919.

7 (Above) *A Massey-Harris 12–25 on display at the Ontario Agricultural Museum.*

8 (Below) *Cletrac Model F showing the roller track designed to reduce power losses.*

9 (Above) *International Harvester 10–20.*

10 (Below) *MG2 tracklayer built by Ransomes of Ipswich.*

11 (Above) *An Allis-Chalmers Model U tractor in Great Britain.*

12 (Below) *A 'Ferguson-Brown' tractor built at the David Brown factory.*

13 David Brown VAK 1A of 1946.

INTERNATIONAL HARVESTER 10-20

There was nothing especially exciting or new about the design of the 10–20, and the Farmall, which arrived on the market at about the same time, attracted much more attention. But the 10–20 offered the sturdy reliability and steady pulling power for which many farmers were looking and the tractor was an outstanding success.

Known generally as the McCormick-Deering 10–20 in North America, the new tractor went into production in 1923 and remained in the IH product line until 1942. During this long production run, output totalled almost 216 000, making the 10–20 a success story few manufacturers could equal.

The dumpy styling of the 10–20 was a smaller version of the 15–30 which had arrived on the market 2 years previously. Both tractors used a similar design of four-cylinder petrol/paraffin engine, with replaceable cylinder liners and overhead valves. Both tractors were fitted with a power take-off shaft as standard equipment, a feature which International had pioneered in the USA.

During its long commercial life, the 10–20 was used as the base for making several more specialised versions. In 1924, the first variant arrived in the form of an industrial model with solid rubber tyres for use on smooth surfaces. Another development came in 1928, when a standard 10–20 was equipped with tracks to become the first crawler tractor manufactured by International.

Apart from the addition of inflatable tyres as an optional extra, the basic design of the agricultural model changed very little during almost 20 years. The weight was increased from the initial 3700 lb in 1923 and the gearing was changed to give a faster travel speed. The price of the tractor in the USA also changed very little, rising from $785 in 1923 to $950 in 1939 for the steel-wheel version. This is a price rise of approximately $10 or a little over 1 per cent a year.

A crawler version of the International Harvester 10–20 was introduced in 1928.

VICKERS-AUSSIE

When the Vickers engineering company decided to expand into the tractor industry they chose the Australian market as their main sales target, rather than their home market in Great Britain. When Australian agriculture was expanding in the 1920s and the demand for tractor power was growing, the Vickers organisation was already firmly established in Australia.

Much of the prototype test programme for the new tractor was carried out in Australia. The name 'Aussie' was used to encourage Australian interest in the tractor, and the unusual rear-wheel design was based on an Australian invention to suit Australian conditions.

The rear wheels appear to have been the main sales feature of the new tractor when it was announced in 1925. Each rear wheel was made in three separate sections, with gaps of approximately 3 in between each section at the rim. Scraper bars, fastened to the tractor, protruded between the rim sections. Their function was to prevent soil building up on the rims in sticky conditions so that the steel lugs would continue to grip effectively.

A description of the new tractor in the British journal, *The Implement and Machinery Review*, claimed that the new wheels gave the tractor such a good performance that it opened up 'a new and brighter era for power-farming'. With its special wheels, the Vickers could work easily where an ordinary tractor would be bogged down and where a man on foot would sink into the soft ground 'to the top of his boots'.

Another advantage claimed for the rear-wheel design was reduced soil compaction because of the wide contact area with the ground.

Exactly why these special wheels were thought to be so suited to Australian conditions, rather than to the heavy clay soils and high rainfall of parts of Great Britain and Europe, is not obvious.

Apart from the wheel design, and the fancy sunshade of the tractor illustrated, the Vickers tractor had a remarkable resemblance to the famous International 15–30. The resemblance is obvious in the styling of the two tractors and there were also similarities in the mechanical design. According to one authority, the Vickers tractor was a copy of the 15–30, following an agreement between the two companies, and some components on the tractors were interchangeable.

The engine in the Vickers tractor, like that in the International, developed 30 hp at 1000 rpm, with a $4\frac{1}{2}$-in bore and 6-in stroke. The gearbox provided three forward ratios and one reverse.

Vickers kept the tractor on the market for about 5 years, although the name 'Aussie' was dropped at an early stage of production. Two of the tractors exported from Britain to Australia more than half a century ago were bought back from their owners in 1978, to be shipped back to Great Britain and restored by Vickers.

CATERPILLAR TWENTY

One of the most significant mergers in American tractor history produced the Caterpillar Tractor Company in 1925. The merger brought together the Best and Holt organisations which had both been in the forefront of crawler-tractor development.

After the merger, the newly formed company increased its development programme to introduce a more comprehensive model range. The 2-ton, 5-ton and 10-ton models were all announced in 1925, with replacements and additions to the product line following at frequent intervals.

The Twenty was a medium-sized crawler model which went into production towards the end of 1927 and remained on the market until 1933. Below it in the Caterpillar range were the Fifteen and the small model Ten, which weighed 4300 lb only; both were built from 1929.

Caterpillar built a four-cylinder petrol engine for the Twenty, with $3\frac{3}{4}$-in bore and 5-in stroke, which developed 25 hp on the belt at 1250 rpm. The pulling power at the drawbar in first gear was 4160 lb, according to the original specification sheet, but a 5721-lb maximum was achieved in the Nebraska Test.

Power was transmitted through a flywheel clutch and a gearbox with three forward ratios and a reverse, plus an optional high ratio third gear available for faster travel speeds. The tracks were steered through dry multiple-disc clutches and contracting band brakes.

There was a substantial market for agricultural crawler tractors in the 1920s, both in the USA and also in many export markets, and the Twenty was designed mainly for farm use.

Overseas demand for the Twenty helped to establish the Caterpillar Company in the export business, including substantial sales to British farms. The medium power and compact size of the tractor helped to encourage interest in the Twenty in areas of smaller farms. The overall width of the tractor was less than 5 ft, with a length of less than 9 ft. Overall weight was registered as 7822 lb in the Nebraska Test report; this was spread over 1090 in^2 of ground contact area with standard 10-in track shoes fitted.

Left. *Vickers tractor with patented rear-wheel design and sunshade.*

Caterpillar Twenties working on snow clearance in the USA and hill land improvement in Wales.

FOWLER GYROTILLER

For a while it seemed that the Gyrotiller would be the product which could restore the Fowler company to the position of prestige and prosperity it had held in the days of steam. Although the machine proved to be only a short-lived success, it was nevertheless an interesting attempt to introduce powered tillage as a replacement in some conditions for the mouldboard plough.

John Fowler was born in Wiltshire in 1826. After a brief period as an apprentice to a corn merchant, he travelled to the north of England where his interest in engineering developed. The 1840s and 1850s in Great Britain were a period when there was considerable interest in the idea of using steam power to cultivate the land. Steam had already transformed transport and industry in Great Britain and there were many Victorians who believed that steam could bring a similar transformation to agriculture.

The first real progress with steam on the land was made by John Fowler, who used a steam engine to provide the power for the system of mole drainage he had developed. This led to experiments with cable-ploughing systems, in which the heavy engine remained on the headland, powering a windlass or winding drum, with a cable pulling the plough to and fro across the field.

Although John Fowler died in 1864, when he was only 38 years old, the company he had established at Leeds was already taking the lead in the technical and commercial development of steam ploughing. The extraordinary growth and success of the company is described in fascinating detail in Michael Lane's history of Fowlers, *The Story of the Steam Plough Works*.

As the market for agricultural steam equipment began to shrink, because of the success of the internal combustion engine, Fowlers were forced to look for alternative products to fill their huge factory. There had already been some diversification into

A Fowler Gyrotiller with MAN engine.

railway engines and also some attempts to use paraffin and diesel engines as a replacement for steam to operate cable-ploughing equipment, but there was renewed urgency to find new products and new markets in the 1920s as the company's financial position was deteriorating.

C.H. Fowler, managing director of the company between 1919 and 1932, met the inventor of the Gyrotiller during a business visit to sugar estates in the Caribbean islands. In Puerto Rico, Charles Fowler was introduced to an American, Norman Storey, who managed an estate and who had recently taken out a Cuban patent covering some of the ideas he was developing in the Gyrotiller. The meeting took place in 1924 and resulted in an agreement for the Fowler company to manufacture the Gyrotiller on an exclusive basis. The first Fowler Gyrotiller was built in Leeds in 1927.

Norman Storey's machine had been developed primarily for the deep cultivations required for sugar-cane production. Two horizontal rotors, powered by a large engine, carried tines which could cultivate undisturbed soil to a depth of 20 in. The tines broke up compacted soil without inverting it and could produce a tilth ready for planting on clean land in a single operation.

Production at Leeds began with a petrol engine. This was a Ricardo design, developing 225 hp and with a thirst for up to 14 gallons of fuel an hour when working at full throttle. The engine had been developed originally as the power unit for a British Army tank and Fowler held a licence to manufacture it. This early version of the Gyrotiller weighed more than 23 ton and had a work rate of about 1 acre an hour with a tillage width of 10 ft. The weight was spread by using crawler tracks, with a single front wheel for steering.

More modest diesel engines were used from 1930, starting with a 150-hp diesel engine built in Germany by MAN. Later a 170-hp Fowler diesel was used and also an 80-hp diesel in a smaller model which had a 9-ft tillage width. The overall length of the big Gyrotillers, including the rear-mounted cultivator, was 26 ft. Both versions had three forward gears for cultivating and a faster gear – two for the smaller model – for road travel.

All the Gyrotillers produced in the early years were exported, with the Caribbean area as the principle market. In 1932, interest began to develop in Great Britain and a considerable demand developed in the main arable areas where contractors were the principal customers. There was also a good prospect for large-scale sales to the USSR, but this market failed to developed after a single machine was delivered there for evaluation.

Production expanded in the mid-1930s, as good reports on the effectiveness of the Gyrotiller were received from the sugar estates and from British owners. Smaller versions of the cultivator were made to fit the 30-hp and 40-hp Fowler crawler tractors and there was a costly programme of demonstrations to ·increase demand and find new markets.

At its peak, the Gyrotiller accounted for almost 50 per cent of Fowler turnover. In many ways it was an ideal product for a company with immense experience of building big, heavy machines to work on the land, selling mainly in areas where Fowler steam tackle had been so successful.

Problems arose while the demand for the Gyrotiller was still expanding. Some of the problems were mechanical and included difficulties with some of the diesel engines. There were also financial problems facing the sugar-producers in the 1930s, which put the high cost of the machines beyond the reach of some estates. A third difficulty was the damage the Gyrotiller could do to the soil when the machine was misused, particularly when cultivating too deeply and too finely in heavy land.

Production ended in 1937, but some of the machines were still working in Great Britain 30 years later.

HOWARD DH22

Australian farmers were among the first to realise the advantages of tractor power and quickly formed a valuable export market for manufacturers in Great Britain and the USA. Australia's own tractor industry developed more slowly and the first substantial success with a tractor designed and built in Australia did not appear until 1928, when the Howard DH22 went into production.

Arthur 'Cliff' Howard, the man who designed the new tractor, was brought up on a farm at Gilgandra, New South Wales. There he began experimenting with powered tillage and, in 1920, he completed his first self-propelled rotary cultivator, with a 60-hp Buda engine and a 15-ft bladed rotor.

In 1922 he formed a company, Austral Auto Cultivators, to build small pedestrian-controlled rotary cultivators and attachments for the Fordson tractor. Later a factory was opened at Northmead, New South Wales, where the DH22 was manufactured.

Although the DH22 could be used as a conventional tractor, it was designed primarily as a cultivation unit with a bladed rotary cultivator attached to the rear. The rotor was powered from the tractor engine through a chain drive from a rear power take-off. The transmission was protected by a slip clutch and there was a choice of sprockets to adjust the rotor speed in order to produce a coarse or fine tilth. The working depth of the blades was controlled by a depth-limiting skid which was fully adjustable.

Mr Howard designed the complete tractor and cultivator unit, including the engine, which was a four-cylinder, overhead-valve unit, with magneto ignition. The engine was rated at 22 hp at 1350 rpm.

Two versions of the tractor were manufactured. The sugar-cane model had a working width of 38 in, which was just wide enough to cover the wheel tracks. On this version, the blades were 8 in long. The standard model, for orchard and field work, according to the instruction book, could be equipped with 4-ft or 5-ft rotors, depending upon soil conditions. Standard on this model were 6-in blades.

Production of DH22 tractors continued for about 20 years. During this time several improvements, including a power lift for the rotary cultivator unit and rubber-tyred wheels, were made to the basic design.

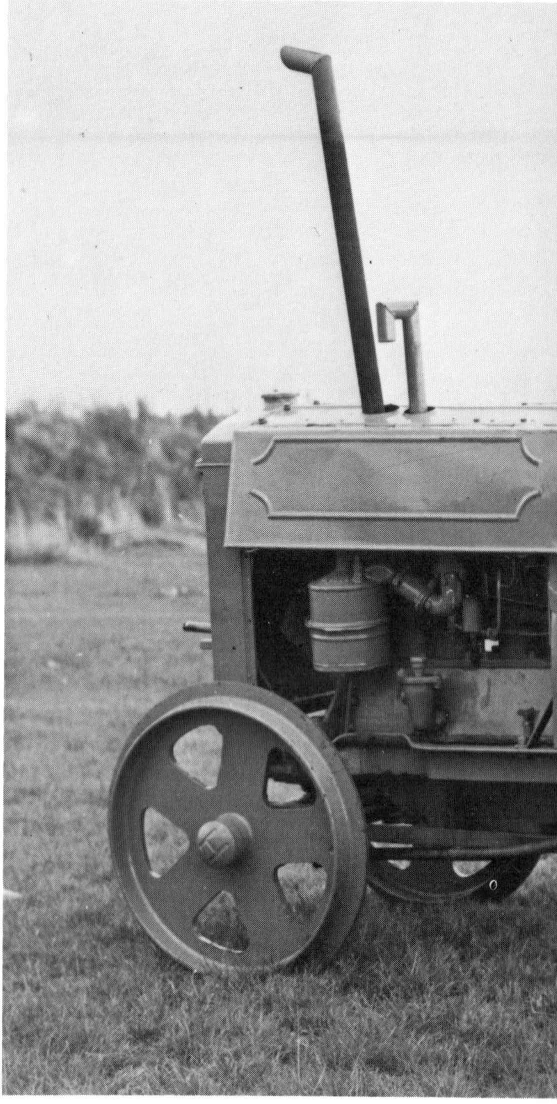

The Howard tractor achieved a considerable commercial success, and there was a significant export demand, especially for the sugar-cane version which was shipped to South Africa and the Caribbean area. One of the standard model tractors was sent to England where it was shown at the 1931 Royal Show – almost certainly the only attempt to launch an Australian tractor on the British market.

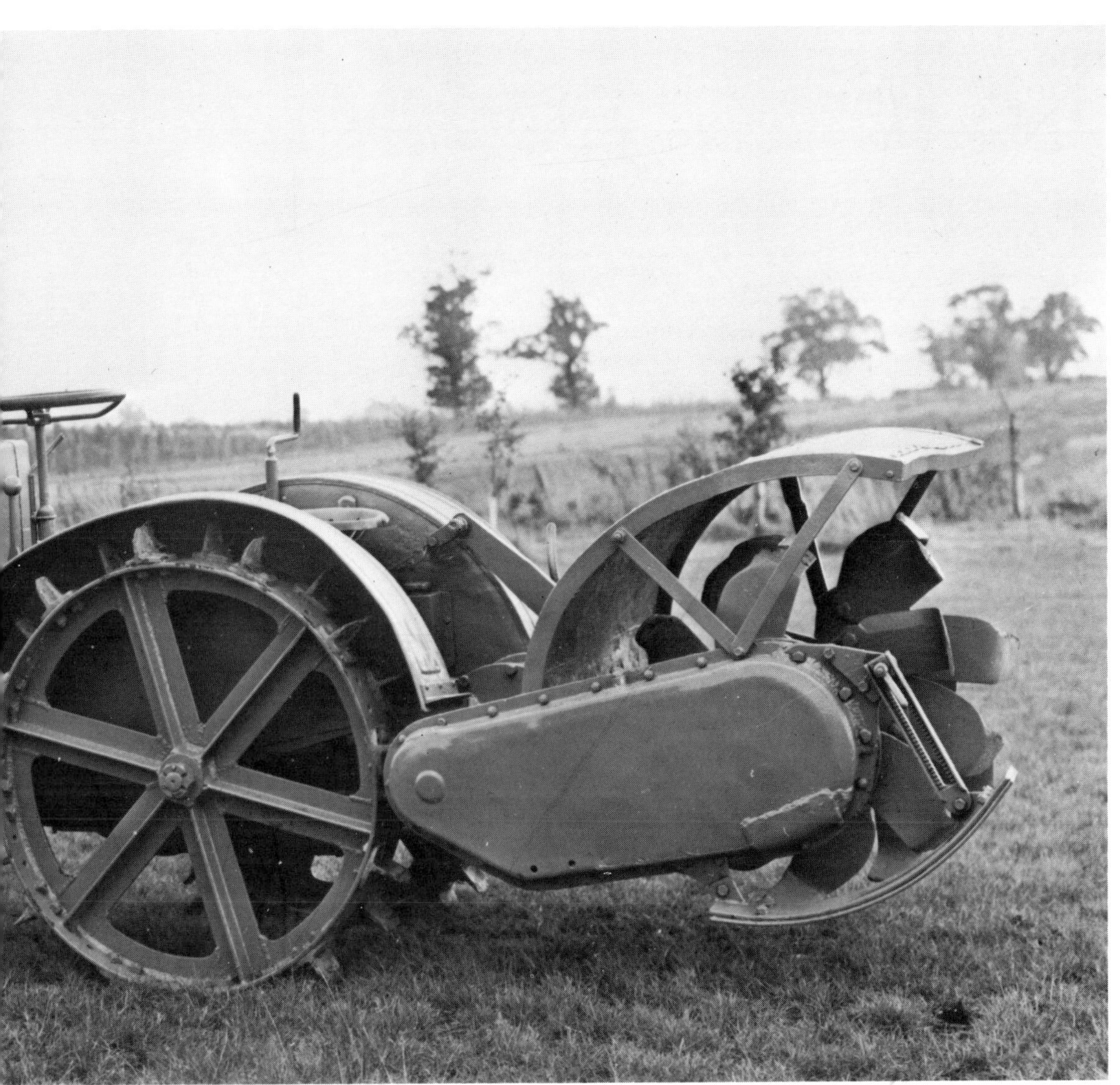

Howard DH22 tractor with rotary hoe attachment.

CASE MODEL C

Case made their first significant impact on the British market in 1930, with the C and L models which had gone into production in the USA in the previous year.

The Model C was particularly suited to British farm conditions, and this helped to encourage interest. The main reason for the extra attention, however, was the performance achieved by the C and L tractors in the 1930 World Tractor Trials.

There were 33 entries for the trials, with tractors brought in from most of the important European manufacturers, as well as a strong representation of American makes. The Trials, which were organised by the Royal Agricultural Society in conjunction with Oxford University, were on an ambitious scale, with the test programme spread over almost 2 months and followed by a large public demonstration.

Case decided to use the Trials as an opportunity to gain publicity for their new tractors. To some extent this must have been a gamble with tractors which had done little work under British conditions and which had been designed primarily for the American market.

Fortunately for Case, the gamble paid off. In the section for wheeled tractors operating on paraffin, the Model C returned the best figures for economy, with the lowest fuel cost to produce its rated load on the drawbar, and equal best for fuel cost on belt work. The test results were described as 'excellent' in the official report. As the Model L had also performed well, Case were quick to feature their success in full-page advertisements in the farming press.

In the maximum load tests at the Oxford Trials, the Model C recorded a highest figure of 29.8 hp on the belt and 21.9 hp at the drawbar. In the equivalent tests at Nebraska in 1929, the belt figure was 29.81 hp and 19.6 hp on the drawbar, with the pull figure recorded in first gear in both tests.

A rowcrop version of the Model C, with a narrow front wheel arrangement, was announced in 1929 and known as the CC. There was also a CI version for industrial work and a CO model for orchard work.

Case Model C.

ALLIS-CHALMERS MODEL U

The Model U became the fastest farm tractor in the world and the star performer in a campaign to introduce rubber tyres.

The letter 'U' stood for United, the name by which the tractor was known when it arrived on the market in 1929. The Allis-Chalmers tractor division had joined forces with others in the farm and industrial equipment business to form the United Tractor and Equipment Company of Chicago. The consortium planned to manufacture and market a range of equipment, including a new medium-power tractor. The tractor was the United, made by Allis-Chalmers and using a Continental engine.

Although the consortium ran into difficulties, the tractor survived and Allis-Chalmers took over the marketing as well as the production. They later fitted their own engine, which developed 34 hp on paraffin compared with the 35 hp of the slightly smaller Continental engine using petrol.

At this stage in its history, the Model U was still an ordinary, undistinguished farm tractor. It became fast and famous as the first tractor available with low-pressure rubber tyres, the forerunners of the tyres used on all modern wheeled farm tractors. For years, engineers had looked for a better type of tractor wheel. Steel wheels with lugs or cleats, fitted to most early tractors, gripped in the field but caused damage on roads and on pasture. Solid rubber tyres and high-pressure truck-type tyres had been tried and neither provided adequate grip for field work. While the problem existed, the value of a tractor in many situations was limited.

The Allis-Chalmers Model U tractor, used for some of the original tests with tyres, on a farm near Waukesha, Wisconsin.

The Allis-Chalmers racing team at speed.

The breakthrough came in 1932, when tests showed that a rubber tyre with a low inflation pressure would give adhesion in the field and avoid problems on the road. Model U tractors were used in the tests, fitted at first with aircraft tyres, which were the most suitable available at the time. The inflation pressure of the driving wheel tyres was 15 psi, which allowed enough flexibility in the walls for the rubber to mould itself to uneven surfaces.

Field scale trials were started in April 1932, on a Model U belonging to a dairy farmer near Waukesha, Wisconsin. Later that year, the first rubber-tyred Model U was sold after a demonstration to a farmer near Dodge City, Kansas.

Although there were genuine advantages, the rubber tyres were greeted with scepticism. Many farmers believed rubber could not be tough enough to stand up to farm conditions and there was concern about the risk of punctures. Allis-Chalmers realised that a really effective publicity campaign would be needed to help sell the benefits of the new tyres.

The theme of the campaign was speed. Apart from their other advantages, the tyres allowed a faster travel speed on the road and this was used to attract attention. While Model U tractors with steel wheels were equipped with a three-ratio gearbox, giving a top speed of about 3.5 mph, a high fourth ratio was supplied on tractors with rubber tyres, giving up to 15 mph travel speed. Advertisements in the farming press carried the slogan, '5 miles an hour on the plow ... 15 miles an hour on the road!'

Meanwhile a secret project was being developed – a Model U especially tuned and equipped for speed. The first public demonstration was at the 1933 Milwaukee Fair. The tractor was shown working with a plough and later it was driven on to the race track. Frank Brisko, a locally successful racing driver, completed a circuit at an average speed of 35.4 mph, which was claimed as a world record for farm tractors.

The high-speed tractor created a sensation and the decision was taken to develop the speed campaign. Allis-Chalmers brought in more Model Us for tuning in order to build up a racing team. Throughout the summer and autumn of 1933, the tractors became an attraction at State fairs and other events, with some of the most famous racing drivers at the wheel, including Ab Jenkins, Lou Meyer and Barney Oldfield. At Dallas, Texas, Oldfield covered a measured mile at 64.28 mph, with American Automobile Assocation officials present to confirm it as a new world record. A few weeks later, Ab Jenkins pushed the record up to 67 mph on the Utah salt flats.

Allis-Chalmers tractor dealers welcomed the speed campaign. Some are said to have done their own engine-tuning in order to exceed the speed limit of 15 mph for farm tractors on public roads. Apparently it was worth a fine in order to gain local publicity for the tractor.

The campaign succeeded. About one million people saw the tractors perform in 1933 and there was widespread press coverage. By 1937, almost 50 per cent of new tractors sold in the USA had rubber tyres.

An Allis-Chalmers tractor at the end of a road run to Chicago in 1933.

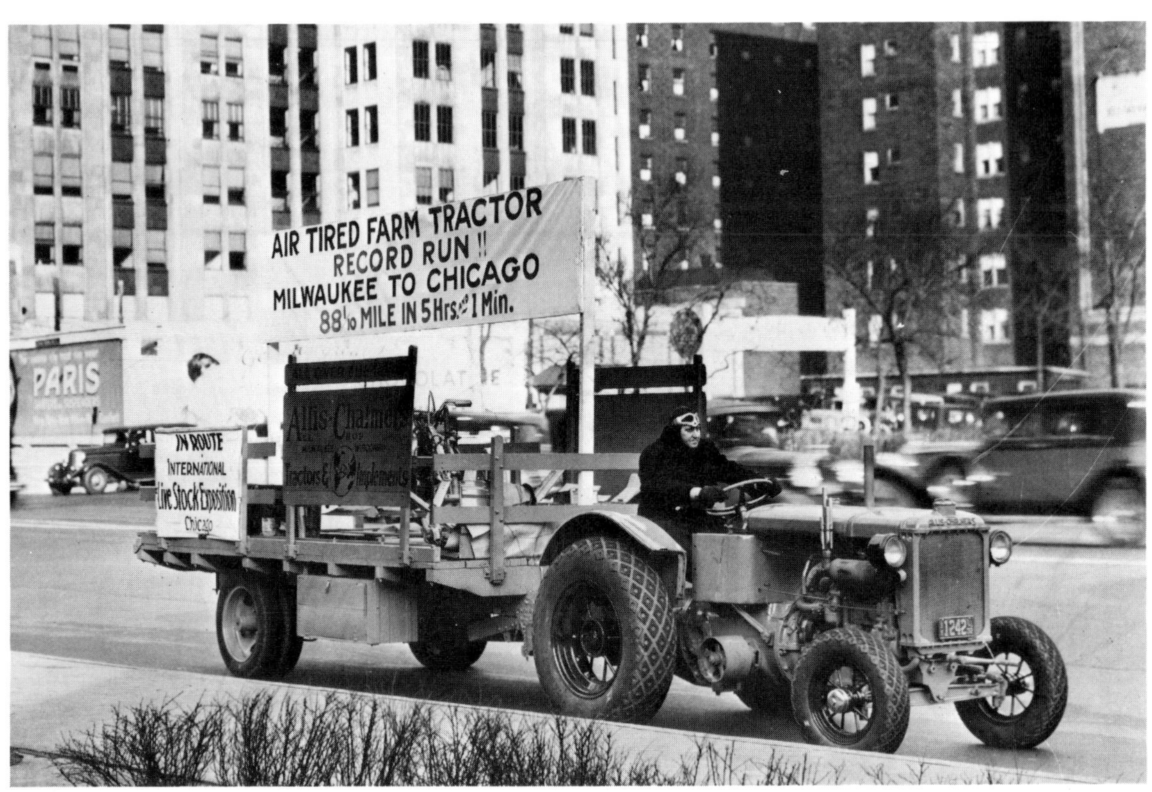

MASSEY-HARRIS GENERAL PURPOSE

Massey-Harris purchased the tractor plant at Racine, Wisconsin, when they bought the J.I. Case Plow Works company in 1928. The company cost Massey-Harris $1.3 million in cash, plus a further $1.1 million in the form of guaranteed bonds. Some of this investment was quickly recovered when Massey-Harris sold their newly acquired right to the Case name to the other Case company, the J.I Case Threshing Machine Company, also based at Racine but not connected with the Plow Works.

It is unlikely that the Canadian Massey-Harris company had much interest in the Case name anyway. The objective in buying the company was to take over the production and marketing of the tractor made by Case in order to strengthen the Massey-Harris position in the North American tractor market. The Racine investment also gave Massey-Harris a much stronger base in the USA.

The first tractor designed by Massey-Harris and manufactured by the company at Racine was the General Purpose. This tractor arrived on the market in 1930 as an unconventional and technically complicated start to the Massey-Harris venture into tractor development.

Four-wheel drive was not new in 1930, but it had not made much commercial impact. To the complexities of four-wheel drive, Massey-Harris added the further complications of four alternative wheel spacings for rowcrop work, plus an oscillating rear axle to cope with uneven surfaces. They also added a system of skid steering, with separate handbrakes to control the wheels on each side enabling the tractor to turn in a 6-ft radius.

Massey-Harris bought in a Hercules four-cylinder engine to power the General Purpose. This developed 22 hp at the belt from a 1200-rpm-rated engine speed. The three-speed gearbox allowed a maximum forward speed of 4 mph and the drawbar performance was 15 hp.

The decision to produce such a technically advanced design proved to be expensive. The development costs for the tractor were inevitably high and these were followed by costly after-sales support to deal with teething troubles in the field. Demand for the tractor proved disappointing. In some ways it was simply too advanced for the market, which favoured simple, low-cost tractors. There was little evidence in the early 1930s of the advantages of four-wheel drive, so that farmers who were prepared to pay extra for improved pulling efficiency were more likely to choose a tracklayer.

Attempts to introduce the General Purpose on to the British market met curious interest but few sales. Improvements to the tractor were made in 1936 when a new version, the 4 WD, was announced. But the improvements and the new name were insufficient to stimulate demand and the tractor

was withdrawn from the market in about 1937.

The General Purpose was an attempt to provide the farmer with a better tractor. Commercially it was a failure, but technically it was one of the most advanced and interesting tractors of its time.

The Massey-Harris four-wheel drive tractor.

MINNEAPOLIS-MOLINE MT

Although the tractor market in the USA tended to expand during the 1920s, there was a big reduction in the number of manufacturers. Some companies simply went out of business or turned to other products as the competition for sales made profits more difficult to achieve. Some of the smaller independent manufacturers chose to join forces in mergers in order to survive against the big companies which dominated the market.

An important merger took place in 1929 when the Minneapolis-Moline Power Implement Company was formed, with its headquarters at Minneapolis, Minnesota. The three companies which gave up their independence in the merger were the Minneapolis Threshing Machine Company of Hopkins, Minnesota, the Minneapolis Steel and Machinery Company of Minneapolis and the Moline Implement Company of Moline, Illinois. The new company offered a wide range of farm equipment plus the Minneapolis tractor range which had been built at Hopkins and the Twin City tractors from Minneapolis. The Moline concern had gone out of the tractor business in 1923.

The new grouping provided an opportunity to rationalise production amongst the three factories. The Twin City tractor range was selected to lead the M-M attack on the market, with production remaining where it had been before the merger, at the Minneapolis factory.

At first the new group simply marketed the Twin City tractors with the Minneapolis-Moline name added. Later, as new models were introduced, the Twin City name was given reduced prominence. One of the first of a wide range of new models announced by the M-M company was the Universal model MT, a rowcrop tractor manufactured between 1931 and 1935.

Like other rowcrop tractors, the MT was designed to take a wide range of mounted implements. These were made at the Moline factory, giving the basis for major expansion into rowcrop farming.

A petrol-paraffin engine made by M-M provided 13 hp at the drawbar and 25 hp at the belt. The engine was a four-cylinder unit with $4\frac{1}{4}$-in stroke and 5-in bore, a design which remained in production, with some modifications, long after the MT had reached the end of its production run.

Production models of the MT were fitted with steel wheels, with a distinctive dished shape for the twin front wheels. The tractor illustrated, preserved at the Greenfield Village and Henry Ford Museum, Dearborn, was equipped with rubber tyres by a previous owner.

Minneapolis-Moline MT tractor in the Henry Ford Museum, Dearborn.

SUPER LANDINI

The history of the Landini company in Italy started in 1884, when Giovanni Landini left the blacksmith's shop where he had been employed to open his own blacksmith's business in Fabbrico. The business was a success and Landini progressed from repair work to building simple equipment for local farms.

More progress came in 1911 when Landini built his first portable steam engine. In 1917, he began to experiment with semi-diesel engines, building some for industrial use. Later he planned to use one of his engines to power a tractor, but he died in 1925 before the first prototype tractor had been completed.

There were three sons to take over the business and they continued with the tractor project, which was completed in about 1925. Later, the first production tractors were built, using a 40-hp semi-diesel engine. This type of engine, two-stroke and with a single cylinder, was popular in some European countries, especially Germany, in the period between World Wars 1 and 2. Landini continued to use the semi-diesel until 1957.

Following some success with their first attempt to enter the tractor market, the Landini brothers began to design a much more ambitious new model. This was the Super Landini, which arrived on the market in 1934 as the most powerful tractor on the Italian market and one of the largest tractors to be built with a semi-diesel engine.

The engine for the Super tractor was rated at 50 hp, which was developed at 650 rpm. Semi-diesels are not noted for their efficiency and the single cylinder of the Super Landini engine had a capacity of 12.2 litres. A modern high-speed diesel engine would probably develop the same amount of power from a total capacity of less than 3 litres. Power from the big engine was transmitted through a gearbox with three forward ratios and a reverse.

Landini tractor sales were still restricted mainly to the Italian market, with little export business, and this limited the demand for such a large, expensive tractor as the Super. Although the Super Landini remained in production until 1939, it was a range of smaller models which helped the company to expand. These included the L25, a new model announced in 1949 with a 4.3-litre semi-diesel engine developing 25 hp. The L25 was produced in a new Landini factory at Como.

As it became increasingly obvious that the multi-cylinder full diesel was taking over from the semi-diesel, the Landini company belatedly signed an agreement with Perkins of Peterborough. This allowed Landini to manufacture Perkins engines in Italy under a licencing arrangement. The first of a new range of diesel tractors was announced in 1956 and the last of the old semi-diesel models was sold in 1961. The Perkins engines allowed Landini to expand their export business, and the association with Perkins also led to the Landini company becoming part of the Massey-Ferguson organisation in 1960.

The Super Landini, Italy's most powerful tractor in 1934.

RANSOMES MG SERIES

The company now known as Ransomes, Sims and Jefferies has been building farm machinery at Ipswich, Suffolk, for almost 2 centuries. Its history, which began in 1789, contains plenty of examples of leadership in the introduction of new ideas.

In 1841, Ransomes demonstrated what was probably the world's first portable steam engine designed for farming. In the following year Ransomes made history again with the first self-propelled agricultural engine. Ransomes manufactured the first commercially successful lawnmower, helped to pioneer a straw-burning system for steam engines and, in 1839, won the first Gold Medal to be awarded by the Royal Agricultural Society of England for their implement display at the Oxford Royal Show. Ransomes has been the leading British plough manufacturer for much of its history and built the first balance plough, which became an essential part of John Fowler's cable-ploughing system.

Ransomes was among the leaders when tractor development was in its early stages in Great Britain. A prototype tractor with a 20-hp engine was built at Ipswich in 1903. It was equipped with a three-ratio gearbox, providing three speeds forward, three in reverse and three speeds on the pulley.

The first commercially successful Ransomes tractor was the MG2, announced in 1936, the first of a series of MG mini crawlers which held a small, specialised share of the tractor market for 30 years.

Total production of MG tractors amounted to approximately 15 000, or an average of 500 a year, including the War years when production was temporarily halted and the post-War period when Ransomes made 1000 tractors a year at the peak of the MG's popularity.

The little tractors were highly unconventional in design. For example, the flywheel incorporated a centrifugal clutch which disengaged when the engine speed fell below 500 rpm. The drive from the clutch was taken through reduction gears to the

Ransomes MG2, the first model in a series of small tracklayers.

two crownwheels and differential gearing. One crownwheel produced a forward gear, while the second gave reverse.

Power from the differential was transmitted by spur gears to the front track sprockets. The tractor was steered by a pair of levers, each acting as a brake on one of the tracks so that the differential acted to speed up the unbraked track. The tracks could be adjusted to suit varying row widths and were rubber-jointed.

In its original form, the MG2 was powered by a Sturmey-Archer 6-hp engine, which was an air-cooled, single-cylinder unit. In the MG5, which replaced the MG2 in 1949, the power output had been raised to 7.25 hp at 2100 rpm, which produced 4.5 hp at the drawbar. The final version, the

MG40, the final version of the MG series, equipped with a Sachs diesel engine.

MG40, was equipped with a 10-hp diesel engine.

With its light weight – 1400 lb for the MG5 – and 74-in length, most of the demand for Ransomes tractors came from nurserymen and market gardeners. The ploughing rate for the MG5 was about 1 acre per 8-hour day, which would have been unacceptable on a larger acreage. There were some other markets for the MGs, including Tanzania, where they were used to scrape salt from the surface of inland salt pans, and also Holland, where the tractors were popular in some areas because they were small enough to be ferried across drainage dykes in small boats.

'FERGUSON-BROWN'

The first Ferguson System tractor was completed in 1933. It was finished in several coats of black gloss paint and was known as the 'Black Tractor'.

Harry Ferguson built the Black Tractor in order to demonstrate the system of three-point linkage and hydraulic controls which he had been developing for about 15 years. In 1933, the Ferguson System was almost complete and ready for production. Harry Ferguson and his team designed and built the Black Tractor to prove the benefits of his method of implement attachment and control to potential manufacturers.

Gears for the Black Tractor were ordered from the David Brown company of Huddersfield, Yorkshire. It was this set of gears which brought Ferguson into contact with David Brown, later Sir David, who became interested in the idea of manufacturing a tractor.

In 1935, Harry Ferguson and David Brown reached an agreement to manufacture and market a Ferguson System tractor. Two separate companies were set up, one controlled by Harry Ferguson, with responsibility for engineering and marketing, and the other controlled by David Brown to manufacture the new tractor. This was a form of agreement which Ferguson later used elsewhere for the Ford 9N and the Ferguson TE20. It was an arrangement which reflected Ferguson's own special interests and it also meant that he did not have to find the heavy investment needed to set up production lines.

David Brown's decision to become a tractor manufacturer was remarkable and far-sighted and it was taken against the advice of his father. At that time the British tractor industry had a discouraging record of commercial failure, with very few companies achieving any long-term stability in spite of a splendid record of innovation.

Tractor production began in 1936, in factory space rented from the gear company. The tractor, which was known as the Ferguson Type A, but more usually referred to as the Ferguson-Brown, was based closely on the design of the Black Tractor. The most noticeable change in the production version was the 'battleship-grey' paint finish. This was the colour Ferguson chose to emphasise the clean, functional styling of the Ferguson-Brown and it remained the Ferguson tractor colour until soon after the sale of Harry Ferguson's interests to the Massey-Harris company.

Engines for the first 500 tractors were supplied by the British Coventry Climax company. The E series 18–20-hp unit was used – a four-cylinder

Ferguson tractors were demonstrated with meticulous care.

petrol or petrol/paraffin engine. This was replaced for the remainder of the Ferguson-Brown production run by a 2010-cc David Brown engine, which developed 20 hp at 1400 rpm. The gearbox provided the three forward ratios and one reverse which Harry Ferguson insisted was correct and the retail price was £224.

The new tractor brought considerable interest wherever it was shown. Demonstrations were arranged with Ferguson's flair for showmanship and concern for detail and the little tractor proved the advantages of the Ferguson System by outperforming heavier, more powerful rivals, especially in difficult conditions.

In spite of superiority of design and the determined marketing efforts, demand for the Ferguson-Brown developed slowly. One barrier was the extra cost of the new tractor. The price of the Ferguson-Brown was nearly twice that of the Fordson, which dominated the market. In addition, special implements had to be purchased to suit the rear linkage of the Ferguson, while the Fordson's conventional drawbar hitch suited the equipment which was generally already on farm.

A stock of unsold tractors began to build up and Ferguson's marketing company was soon facing financial difficulties. The money problems were eased by forming a joint company with David Brown and Harry Ferguson as managing directors, but this arrangement failed to increase demand and did nothing to improve the strained relationship which was developing between Brown and Ferguson.

Much of the friction arose over disagreement on how the company's problems could be resolved. Ferguson's answer was to cut the selling price in order to increase demand. David Brown believed the tractor failed to provide what the customer wanted and he favoured a new model with a more powerful engine and a fourth gear ratio.

The partnership ended early in 1939, after Ferguson had made his successful journey to meet Henry Ford and David Brown had gone ahead with development work on a new tractor without the approval of his partner.

In strictly commercial terms, the first Ferguson production tractor had achieved little success. Only 1350 were built between 1936 and 1939, in spite of an expensive marketing effort. Much more important than the sales figures is the fact that the tractor proved the benefits of the Ferguson System beyond any doubt and was also the beginning of the David Brown tractor company.

FORD 9N

The success of the 9N tractor was reasonably predictable. It was the product of two of the most talented men the tractor industry has known and it arrived at a time when demand for tractors was expanding again after a period of depressed sales.

The two men who worked together were Harry Ferguson and Henry Ford. Ferguson contributed engineering skills and the original thinking which he and his team had already demonstrated in the development of the Ferguson System and the Model A tractor. Henry Ford provided the unrivalled resources and experience in large-scale production which had already produced successes such as the Model T car, the Trimotor aeroplane and the Fordson tractor.

Agreement to manufacture a new tractor followed a demonstration to show Henry Ford the advantages of the three-point linkage and hydraulic controls which Ferguson had developed. Harry Ferguson arranged the demonstration at Dearborn with his usual attention to detail, using a Ferguson A tractor brought from England for the purpose.

Ford was impressed by the tractor and by Harry Ferguson. The agreement suited them both, as Ford was planning to get back into tractor production in the USA, while Ferguson was seeking a manufacturer. The exact terms of the agreement they reached are uncertain as the details were never witnessed or written down, but Ferguson and Ford appear to have had a complete understanding of how they should work together and their partnership began successfully.

One indication that the agreement was working well was the excellent tractor it produced. The 9N was probably the most advanced tractor available, presenting the Ferguson System in neat styling and at low cost.

Another sign that the project was developing successfully was the speed with which it progressed. Ferguson demonstrated his British-built tractor to Ford in October 1938, and this was when the agreement began. By January 1939, three experimental tractors incorporating the Ferguson System had been built and were being tested. In June 1939, production of the new 9N had started, with 500 guests attending the official launch party. During this hectic period, Ferguson had terminated his agreement with David Brown, made a substantial start on the development of a distribution network for the tractor in the USA, arranged the production of some of the implements to suit the new tractor, and had worked closely with the Ferguson and Ford engineers on almost every detail of the new design.

It was the 9N which put the Ferguson System on to the market in big volume.

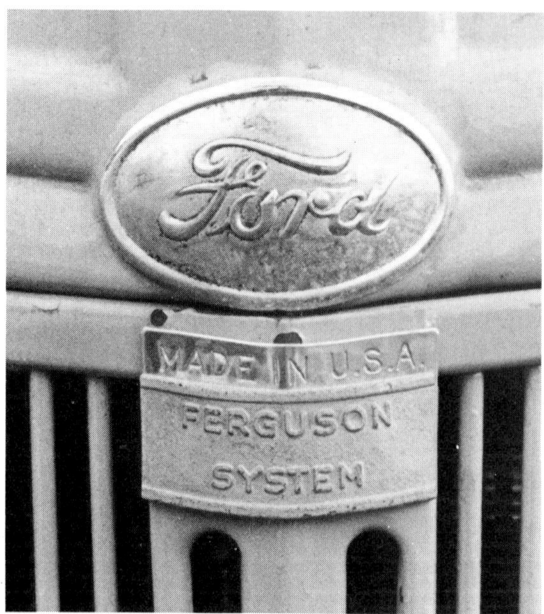

The 9N brought together two famous names—Henry Ford, whose name appeared at the top, and Harry Ferguson, whose name was on the small rectangular plate below.

Under the terms of the agreement, the Ford organisation was responsible for manufacturing the new tractor, while Ferguson and his partners were in charge of marketing. The tractor was sold under the Ford name, but also carried a plate with the words 'Ferguson System'. To most people, then and now, it was known as the Ford-Ferguson.

The 9N engine was a four-cylinder, side-valve unit, built by Ford, and developing up to 23.5 hp at 2,000 rpm when tested at Nebraska. There was a three-ratio gearbox and a maximum drawbar pull of 16.3 hp in the Nebraska Rated Load Test. But the outstanding feature of the 9N was the three-point linkage and hydraulic system, available in a mass-produced tractor for the first time.

More than 300 000 9N tractors were built before the Ford-Ferguson agreement ended in bitterness in 1947.

DAVID BROWN VAK 1 AND VAK 1A

The VAK 1 was the first tractor designed, built and marketed by the David Brown company. Previously David Brown had also manufactured the first Ferguson production tractor, but this tractor had been designed by Harry Ferguson and marketed by a company Ferguson controlled.

Sales of the Ferguson tractor had been lower than expected and David Brown – later Sir David – believed design changes were necessary to encourage demand. He wanted more engine power and a fourth gear ratio. Although Harry Ferguson had firmly resisted the changes, David Brown engineers were working on them before the Ferguson-Brown partnership ended in January 1939.

The end of the partnership left David Brown free to put his own ideas into production. The result was the VAK 1, painted a bright shade of red which the company called 'hunting pink' and launched with considerable success at the 1939 Royal Show at Windsor. The tractor was manufactured in a new factory at Meltham near Huddersfield, Yorkshire, which is still the home of David Brown tractors.

Extra power for the new tractor came from a 2.5-litre engine instead of the 2-litre engine used for the Ferguson. This gave the new tractor 35 hp at 2000 rpm, with 4.7:1 compression ratio. Other improvements compared to the previous Ferguson specification included four forward gears instead of three, and a power take-off, which was designed into the tractor instead of as a bolt-on extra. A hydraulic three-point linkage was part of the VAK design, but with a depth-limiting wheel instead of the hydraulic depth control which was protected by Ferguson patents.

In spite of the extra power and other improvements, including a more comfortable seat and sleeker styling, the price of the VAK 1 in 1939 was £220, which was slightly lower than the original price of the Ferguson model it replaced. The new tractor was also compact in relation to its power and the unballasted weight was only 3585 lb.

Reaction to the new tractor was favourable, but soon after full-scale production had started, World War 1 began and disrupted the flow of tractors from the factory. During the War, the David Brown factory was working on Government contracts. These included tracklaying tractors for the British Army and the VIG series tractors built for the Air Ministry and specially designed for towing aircraft, fuel bowsers and bomb carriers for the Royal Air Force.

Because of the wartime restrictions, total production of VAK 1 tractors was only 5350 when, in 1945, a new version, the VAK 1A, was introduced.

Differences between the old and the new versions of the VAK were a number of detail improvements introduced on the 1A as a result of experience gained with the 1939 model. Among the more important changes were an improved engine lubrication system, a more accurate governor to control engine speed and an automatic hot-spot to allow a faster changeover from petrol to paraffin.

Production of the VAK 1A ended in 1947, when 3500 had been built. The replacement was the highly successful Cropmaster, a development from the VAK tractors but with significant improvements which helped to bring David Brown into prominence as an international manufacturer.

Production of VAK tractors was held up during the war to allow manufacture of specials such as the David Brown aircraft tug and an experimental crawler model based on the VAK design.

MINNEAPOLIS-MOLINE UDLX

One of the more recent developments in tractor design has been an improvement in conditions for the driver. More legislation is being introduced to protect the driver's safety and hearing and manufacturers are competing to offer extra comfort and convenience.

In the 1930s, life for the man on the tractor seat was much harsher. Attempts to interest farmers in equipment such as padded seats, electric starters and cabs to keep out the rain met widespread sales resistance, although these items were often specified on the industrial versions of farm tractors.

Minneapolis-Moline introduced their UDLX model in 1938 in an effort to develop a new market for a tractor with very high standards of safety, comfort, speed and style. The design project started in 1935 and was supported by market research which suggested a worthwhile market for a tractor of this type.

Only about 150 UDLX tractors were sold during the 2 years they were on the market and this total included a significant number sold as industrial tractors. Although the market research was proved wrong, this was primarily because the UDLX had so many features which were 30 or 40 years ahead of their time.

The cab of the UDLX was designed for safety as well as for weather protection. It was made of steel, strong enough to provide some protection from an impact, and with insulation inside to keep down the noise level and help control the temperature. Other safety features included three front-windscreen wipers, safety glass in all windows, front and rear lights, stop lights and self-energising Bendix brakes.

For the fortunate few who spent their working hours at the wheel of a UDLX, there was a range of luxury fittings which only the most expensive cars could match. The two seats were fitted with back rests and were deeply upholstered. Standard equipment included a fitted radio with telescopic aerial, cigar lighter, glove compartment, sun vizor, rear-view mirror, electric clock, heater, hot air defrosting vents, instrument panel light, roof light and rubber mats on the floor. The UDLX was probably the only tractor in the 1930s with a fitted ashtray.

Controls included a foot throttle which would take the tractor up to 40 mph in fifth ratio, a horn, electric starter button, foot-operated dip switch for the headlights and change-on-the-move gears. The front windows opened, the cab door – at the rear of the tractor – could be locked, there was a spot light on the roof, and front and rear bumper bars. The styling reflected the latest trends from the Detroit automobile industry, where streamlining was becoming fashionable.

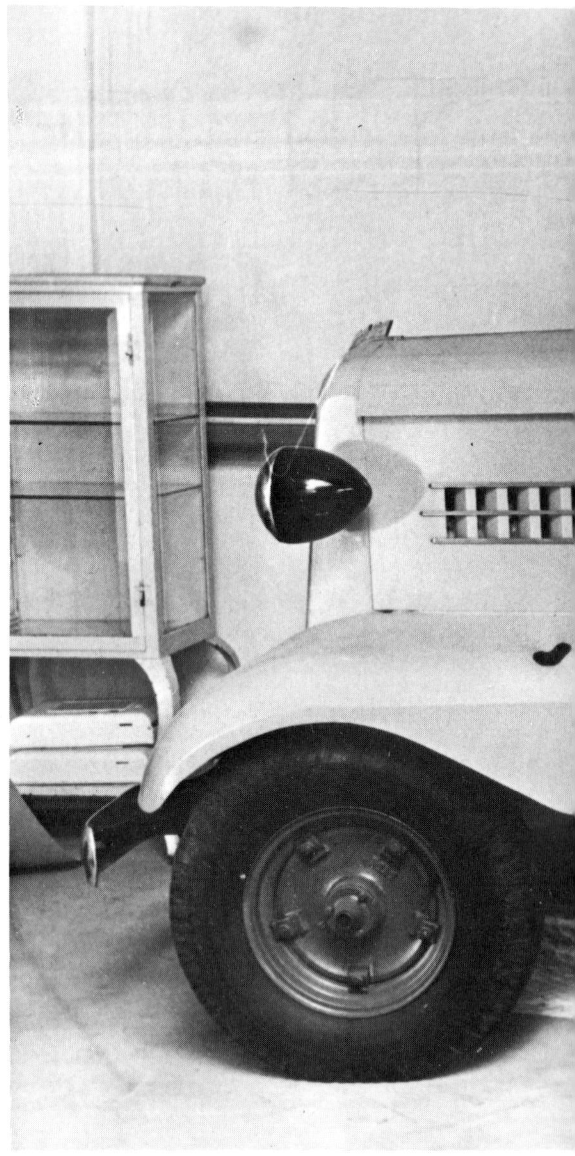

Inevitably, the cost of providing such a com-

prehensive and luxurious specification was high. The 1938 ex-works price for the UDLX was $2155, which was twice the price of an ordinary 40-hp to 50-hp tractor, and more than most farmers were willing to pay.

Although the UDLX attracted few customers and made little impact on the market, it deserves recognition as the first tractor designed with real concern for the driver.

The luxurious Minneapolis-Moline UDLX preserved at Dodge City, Kansas.

JOHN DEERE MODEL B (STYLED)

The success of John Deere tractors in the North American market during the 1920s and early 1930s was due mainly to their performance. The twin-cylinder horizontal engine, which powered most John Deere tractors at that time, had a remarkable reputation for simplicity and reliability, and made a big contribution to the John Deere success story.

But even enthusiastic John Deere supporters might have agreed that the tractor range was beginning to look old-fashioned in the mid-1930s, at a time when styling was becoming increasingly significant. The appearance of John Deere tractors changed very little between 1923 and 1937, while most of their rivals were adopting more rounded lines which the car industry was making fashionable.

To deal with the problem, John Deere commissioned Henry Dreyfuss, one of the leading industrial designers working in the USA at that time, to create a completely new style for the company's tractor range. The first of the 'styled' models, the A and B, were announced in 1938, followed by the D and H in 1939 and the G in 1942.

The new look immediately scored an outstanding success and John Deere tractors began to sell because of their looks as well as their performance. The Dreyfuss styling was still evident in the John Deere range 20 years after it was introduced and still looked stylish.

The model B had first appeared in 1935. It was designed with the smaller acreage rowcrop farmer in mind and, in its original form, was rated at only 9.3 hp at the drawbar. The engine, as usual, was a twin-cylinder horizontal unit.

There were several variants of the basic model B, including the BO for orchard work, the BN, with a narrow, single-wheel front end, and the BR, which was the regular fixed-axle model. John Deere kept the model B in production until 1952, with improvements introduced at frequent intervals, and increased power outputs.

The styled Model B from John Deere.

CLETRAC GENERAL

After more than 20 years of specialising in crawler tractors, the Cleveland Tractor Company announced their first wheel-type model in 1939. This was called the General GG, with an equivalent crawler model known simply as the HG.

The Cleveland company had been using Hercules engines in some of their Cletrac crawler models for many years and Hercules was chosen to supply the power unit for the GG and HG tractors. For both models, this was the IXA engine, a four-cylinder unit with 3-in bore and 4-in stroke, developing 19 hp on the belt at 1700 rpm. Later a slightly more powerful engine, with the bore increased to $3\frac{1}{8}$ in, was used to bring the power at the belt up to 22 hp.

The General was a rowcrop tractor with a tricycle wheel arrangement and was designed to take a range of mid-mounted implements. The B.F. Avery Company of Louisville, Kentucky, produced implements for the General and, in 1941, Avery also took over the production of the tractor.

This arrangement was logical for the Avery company as it meant that they were able to offer a tractor with the implements they manufactured. The tractor was given a modest face-lift, including a more rounded front-end styling, and was marketed as the Avery Model A. Later the power was again increased by fitting the Hercules IXB engine.

Following the deal with Avery, the Cleveland company's brief involvement in the wheeled tractor market came to an end. The decision was probably prompted by the need to concentrate resources on the production of crawler tractors to meet a US Government contract. A large number of Cletracs were supplied during World War 1 for towing Flying Fortress aircraft and this was much more important to the company than building General tractors for the highly competitive rowcrop market.

Before the War ended, the Cleveland Tractor Company had been sold to the Oliver Corporation in a deal which took place in 1944. In 1960, the Oliver company was sold to the White Motors Corporation where the Oliver brand name was retained for almost 20 years. Meanwhile the Cletrac name and product line had quietly disappeared, hastened on their way by the success of four-wheel-drive tractors and shrinking demand for agricultural tracklayers.

The General, the first wheeled tractor built by Cletrac.

MINNEAPOLIS-MOLINE GT

For many years, the GT series were the biggest tractors in the Minneapolis-Moline range. They were rated as 49 hp, but, in the Nebraska Maximum Load Tests, the peak output on the belt was 55 hp. The weight recorded at Nebraska was 9445 lb, making the GT one of the heaviest wheeled tractors on the American market during the early 1940s.

Minneapolis-Moline launched the GT as a 'five-plow' tractor in 1939 as part of a batch of new models to modernise their range. The new models offered more up-to-date styling and were better equipped than most previous M-M tractors. But the power unit for the GT had first appeared several years previously and was a solid, but rather old-fashioned, design. The engine developed its power at a leisurely 1075 rpm from four cylinders with $4\frac{5}{8}$-bore and 6-in stroke and was available initially as a petrol version only.

Production of the GT continued during the War years, in spite of converting much of the M-M factory capacity to making military equipment, including Jeeps for the American Army. Some of the wartime GT tractors were exported to Great Britain to help provide the extra power required to increase food production. With up to 5000 lb of drawbar pull available, the GTs provided a new experience of power for British farmers accustomed to the more modest performance of Fordsons and early Ferguson tractors.

After World War 1, Minneapolis-Moline continued to offer basically the same petrol engine in various replacements for the GT, including the G and GB models, but with additional power developed. The model G of 1950 was rated as 58 hp and the GB tested 5 years later developed 65.5 hp.

Minneapolis-Moline GT.

INTERNATIONAL HARVESTER W6 AND FARMALL M

These were among the most popular tractors in the American 'three-plow' category during World War 2. The W6 version was announced in 1940 as part of a batch of new International models which also included the smaller W4 and the big 50-hp W9. The M was in production a year earlier when the Farmall range was brought up to date.

Although the tractors looked different, they shared the same engine type. This was a four-cylinder unit with $3\frac{7}{8}$-in stroke and $5\frac{1}{4}$-in bore. This engine developed a maximum 36.6 hp on petrol and 34.8 hp in the lower compression distillate version, in the Nebraska Two-Hour Maximum Load Tests.

Diesel engines were available for both tractors, which were identified as the WD6 and MD models, from 1940. Although there was little difference in the power output from the diesel and petrol versions, the Nebraska Test results showed lower fuel consumption for the diesel.

Improved versions of both tractors went into production in 1952. These were the Super W6 and Super M, with a 'D' added in each case to denote the diesel version. An extra 10 hp was achieved by increasing the engine bore to 4 in.

The Farmall M became the first tractor to be assembled by International Harvester in England at their Doncaster factory. The factory was already producing a range of IH farm machinery and tractors were added in 1949 when an assembly line was added as part of a factory expansion project.

The British-made IH tractors were identified by the letter 'B', so that the diesel version of the M, first produced at Doncaster in 1952, was known as the BMD. Just to complicate the sequence, the BMD was replaced in 1954 by the BWD6. The first British-built IH crawler tractor was the BTD6, announced in 1953. The BTD6 and the American T6 and TD6 were the tracklaying equivalents of the M and W6 wheeled series.

Farmall Super BMD built at Doncaster, Yorkshire.

McCORMICK
INTERNATIONAL
FARMALL
DIESEL
SUPER
BMD

CASE LA

When Case announced their new Model L tractor in 1929, they were making their first break from the previous cross-mounted engine design which had been a Case speciality for 15 years. The old tractor range had been a success, but the new Model L was even more successful and earned an outstanding reputation for its performance on belt work.

The Model L remained in production until 1939, when it was updated and restyled as the LA, which came on the market in 1940.

Many enthusiasts think the square, functional styling of the Model L and other Case tractors in the 1930s was as distinctive and attractive as any in the industry. The appearance of the LA followed the trend to more rounded outlines, which made the tractor look more fashionable – but also more ordinary.

Beneath the new shape, much of the engineering which had made the L a success was continued unchanged in the LA. The engine was basically the same four-cylinder, overhead valve unit with about 6.6-litre capacity. In the later versions of the Model L, the hp figure in the Nebraska Tests was 47. This was raised to almost 60 hp in the LP gas version of the LA.

Another well-tried feature of the Model L, which survived through the production life of the LA, was the use of two chains to take the drive to the rear axle by means of twin sprockets. Although this was beginning to appear dated, the chains were rugged and trouble-free and rarely required adjustment. The gearbox was conventional, with four forward ratios and a reverse, driven from an overcentre-type clutch controlled by a lever.

Retaining some of the ruggedness of the old Model L, the LA carried the Case company through the years of peak demand for tractors during and after World War 1. In many respects, the design was already dated, but it remained in production until 1955, when a completely new range of tractors was gradually introduced. The new models did away with many of the old-fashioned features of the L and LA, introducing a foot-clutch, live power take-off and an optional dual-range gearbox with eight forward speeds. The styling was again changed to keep abreast of fashion, bringing in built-in headlights and a new 'desert sand' main colour.

Case LA with modern rounded lines, but an old-fashioned chain drive.

FIELD MARSHALL SERIES 1, 2, 3 AND 3A

Marshall wheeled tractors were familiar in many parts of Great Britain during nearly 30 years of production. They also had a distinctively loud sound, as the single-cylinder diesel engines made a resoundingly recognisable exhaust note.

A prototype tractor was built in 1929 and production started in the following year and continued on a modest scale until 1945, when the end of wartime restrictions brought a big increase in demand and an opportunity to expand production.

In June 1945, a new model was announced with more modern styling and some technical improvements. The new tractor was the Field Marshall Series 1, using an engine which was based on the original 1930 design. The engine bore was $6\frac{1}{2}$ in with 9-in stroke, as on previous post-1935 models, but the 1945 version developed 40 hp, an improvement achieved partly by increasing the speed from 700 to 750 rpm.

Another development on the Series 1 was the introduction of cartridge starting for cold weather, which was a welcome improvement on relying on the handle. The new model was also equipped with a differential lock and a transmission brake, a bigger fuel tank and an optional power take-off. Production of Series 1 tractors continued at the rate of 1000 a year until 1947, when the Field Marshall Series 2 was announced.

With the Series 2, which also remained in production for 2 years, Marshall introduced independent rear-wheel brakes and detail improvements which included improved engine-cooling. Field Marshall tractors were proving especially popular with agricultural contractors because they were particularly satisfactory for stationary work and also because of the reputation for reliability which had been established. Marshall recognised and encouraged this by offering a contractors' version with winch and dynamo lighting.

This Field Marshall Series 3 was photographed in the Lebanon destroying a field of hashish.

Field Marshall Series 3 tractors still retained the two-stroke diesel engine, but linked it to a double-ratio gearbox with six forward gears and two reverse. Wheel sizes were increased and the rear-wheel brakes were now operated by foot-pedals instead of the hand-levers of the Series 2. The power take-off was now central.

By this time, demand for the Field Marshall was beginning to fall away from the peak levels of 1948 and 1949. One reason for the declining sales was the arrival of increased competition at the upper end of the power scale, including some tractors with Perkins engines which brought new standards of smooth power and made the slow-revving Marshall engine seem out-of-date.

To help hold off the competition, the Field Marshall Series 3A was brought on to the market in 1952. The arrival of the new model was made obvious by a change of colour from the subdued green of previous Field Marshalls to a bright, eye-catching orange. Engine efficiency was improved by using a different piston design and by raising the fuel-injection pressure. At the rear of the tractor, a special hydraulic implement lift was available for the first time on a Marshall.

The Series 3A remained on the market with declining sales until 1957, when the company finally brought the Field Marshall series to an end. It was also the end for the VF and VFA crawler tractors which had been produced in parallel with the Field Marshalls, and the end of the single-cylinder diesel engine which had helped to keep Marshall in business for nearly 30 years.

FORDSON E27N MAJOR

When the first of the new Fordson Majors came off the Dagenham production line in March 1945, they attracted a great deal of interest. The Model N Fordson was at the end of probably the most successful production run in the history of farm tractors, having built up a remarkable reputation for sturdy reliability and value for money. The replacement of such a familiar and well-liked tractor was a matter of importance.

Reaction to the new tractor was generally favourable. The Fordson Major looked bigger than the old Model N, more powerful and more purposeful. But, beneath the new styling, little had changed. Mechanically, the Fordson Major was basically similar to more than three-quarters of a million F and N Fordsons built since 1917.

One of the few substantial changes introduced with the new model was a spiral bevel and conventional differential in place of the worm drive which Ford had used for almost 30 years. There was also a single-plate wet clutch to replace the previous multi-plate unit.

These changes gave the Major a more efficient transmission and reduced the distinctive noise of the worm drive. They also allowed Ford to tidy up the back-end design of the tractor to make it easier to fit hydraulic lift arms.

The 1945 tractor still used the 1917 engine design, with four cylinders and side valves, but with the extra capacity which had been provided for the Model N and with the rpm raised to 1450 to push the maximum power output to about 30 hp.

Another, more visible, difference was the use of larger-diameter wheels for the Major. This helped to account for the more imposing appearance of the new tractor and also achieved better ground clearance so that underslung equipment could be used.

There were four versions of the Major; the Standard Agricultural model with steel wheels was the cheapest on the price list at a basic £237 in 1945. The Land Utility version offered a slightly better specification, with rubber tyres as standard equipment and independent brakes as an option. The Rowcrop Major could be ordered with either steel or rubbers and was also available with independent brakes. The fourth version was the Industrial model, with electric starting and lighting, high top-gear ratio and a special towing hook as standard equipment.

All three agricultural versions were available with a choice of standard ratios – red spot – or with special low gear ratios – green spot. A further

A Fordson E27N with a Perkins diesel engine.

option, available from 1948, was the Perkins P6 diesel engine. As the Fordson Major gearbox and other transmission components continued to operate reliably with the 45 hp of the P6 engine, they must have been comfortably understressed in the original Fordson F.

Ford offered good value for money, excellent reliability and unrivalled availability of parts, which made the Fordson Major a popular choice for companies offering specialised conversions. Some of the more important of these were the County tracklayer, Roadless half-tracks and the Italian Selene four-wheel drive model. The Fordson skid units were also popular as the basis for various self-propelled machines.

The Fordson Major became an outstandingly popular tractor, with production rising to a peak of more than 50 000 in 1948. This was almost double the output of the Fordson N tractor in the best years of its production at Dagenham. Some of the Major's sales success was achieved overseas, exports accounting at times for more than 75 per cent of production.

A completely new Fordson tractor, the El ADKN New Major, went into production at Dagenham in December 1951 and the last of the E27N tractors was manufactured early in 1952. This marked the end of the E27N Major, and it also marked the end of the original Farkas design, which had survived with surprisingly little modification for more than 34 years and about one million tractors.

HURLIMANN D-100

Switzerland has a small farming industry with little land suitable for arable crop production. In spite of the limited size of the available market, there are several tractor manufacturers based in Switzerland, including the Hurlimann company which celebrated its fiftieth anniversary in 1979.

Hans Hurlimann, who founded the company, used a single-cylinder petrol engine for his first tractor, which was equipped with a power-operated cutter bar for mowing grass. Although his first tractors looked old-fashioned, improvements during the 1930s began to give the company a reputation for advanced design. The improvements included a new model in 1939, which was powered by a four-cylinder, direct-injection diesel engine.

With limited sales potential in the Swiss agricultural industry, Hurlimann looked for other sales outlets for the new diesel tractor. These included exports to other European countries, the Swiss Army – which bought hundreds of Hurlimann tractors for moving field guns – and industrial sales, for which a special model was developed complete with luxury cab.

An improved version of the diesel tractor went into production in 1945. This was the D-100, comprehensively equipped with a two-speed pulley and power take-off, differential lock, hand- and foot-throttles and five forward speeds and a reverse. The diesel engine was rated at 45 hp at 1600 rpm.

The 1939 Hurlimann diesel from Switzerland.

A design feature of the D-100 was the low centre of gravity, which was helped by using 22-in-diameter rear wheels. This was intended to give extra stability on the steep land which is a feature of many Swiss farms. It was also the reason why one of the tractors was imported in about 1949 for a large-scale reclamation project on hill land in Wales.

This, apparently, was the only Hurlimann tractor imported into Great Britain until 1979, when the modern range of Hurlimann diesels was launched on the British market. The 1949 import is preserved in Philip Jenkinson's collection at the Shebbear Farm Museum in Devonshire.

Hurlimann is now a subsidiary of the Italian SAME organisation.

One of the biggest customers for the Hurlimann diesel tractor was the Swiss army.

Hurlimann produced a special industrial version of their diesel tractor, complete with road lights and purpose-built cab.

MASSEY-HARRIS 44 AND 744

When World War 2 ended in 1945, plans were already being made for a new range of Massey-Harris tractors designed to increase the company's share on the North American market.

The new tractors went into production in 1947, resulting in the most comprehensive range in the company's history. They included the little one-plough Pony and the 59-hp model 55, which was more powerful than any previous Massey-Harris tractor. The highly competitive three-plough sector was covered by the model 44 tractor.

Production of the 44 in its various forms totalled more than 90 000 units from the Racine factory in the USA. The best production year was 1955, when almost 20 000 were manufactured. This made the 44 easily the best-selling Massey-Harris tractor in the North American market at that time.

The distinctively-rounded styling of the 44, like that of other tractors in the 1947 range, was based on the appearance of Massey-Harris tractors of 1939. But underneath the curves there were some important changes, including different engines. For the 44, there was a choice of four-cylinder petrol or paraffin engines and a four-cylinder diesel engine which developed 42 hp. The specification was improved further in 1950 when the hydraulic lift system for mounted implements was introduced.

The 44 was the model Massey-Harris chose when they decided to begin tractor production in Great Britain. The decision to build tractors in Great Britain was taken after efforts to link up with the British Nuffield and David Brown companies failed. Massey-Harris added the 44 tractor to the range of farm machinery already being manufactured by their British subsidiary, in order to take advantage of export opportunities to the important British Commonwealth markets.

Production started with a pilot batch of 16 tractors from the Manchester factory in 1948. Full-scale manufacturing began the following year at the new Kilmarnock factory in Scotland, where Massey-Harris had also established their British production lines for combines and balers. The tractor was known at first as the 744PD, and later called the 744D after production had moved to Scotland.

To some extent, the tractor operation in Great

A British-built Massey-Harris 744D with half-track conversion.

MASSEY-
HARRIS
744D
MASSEY-HARRIS

A publicity photograph of the 744D.

Britain was an assembly process, using components imported from Racine. These included the excellent gearbox, which offered five forward ratios at a time when most European tractors still had only three or four. The engine for the 744D was bought in from Perkins of Peterborough. This was the famous P6 diesel, with six cylinders, totalling 288.6 in^3 capacity and developing 46 hp.

Most of the 744s were the standard four-wheel version, but there were also high-clearance models, a rowcrop version with twin front wheels, plus a half-track model equipped with Roadless tracks.

Demand proved to be disappointing, especially in Great Britain. Less than 17 000 were built in 6 years at a time when the Ferguson factory at Banner Lane, Coventry, was manufacturing more than 50 000 tractors a year.

A new version, the 745, was announced in 1954 and was powered by a four-cylinder Perkins engine. Approximately 11 000 were manufactured before production ended in 1958.

FERGUSON TE-20 AND TO-20

The unconventional agreement between Harry Ferguson and Henry Ford produced the immensely successful 9N tractor, but it also caused a number of difficulties for Ferguson which led eventually to the production of the TE-20 in Great Britain and the TO-20 version in the USA.

One of the early problems arose over the manufacturing policy at the Ford factory at Dagenham, Essex. Harry Ferguson expected the English factory to switch from manufacturing the ageing, but still popular, Fordson to manufacturing a Ferguson System tractor soon after World War 2 started. He appeared to believe that this was part of his agreement with Ford.

Ford directors in England decided against a model change, perhaps because this would have been difficult to implement in wartime conditions and perhaps also because of reluctance to work closely with Ferguson. Eventually, Harry Ferguson accepted the situation and, during the War, began to plan alternative arrangements for manufacturing tractors in Great Britain.

Soon after the War ended, Ferguson's plans developed rapidly. In November 1945, he formed a British company to design and distribute a new tractor, while negotiating a manufacturing agreement with the Standard Motor Company. The new tractor was the TE-20 which went into production at the end of 1946 in the Standard factory at Banner Lane, Coventry.

The TE-20 was basically similar to the Ford 9N tractor, which was still being manufactured in the USA under the original agreement. Important differences between the two included a more powerful overhead valve engine and a fourth forward ratio in the new Ferguson.

Demand for the new tractor developed rapidly in Great Britain and there was also a strong export demand. Meanwhile, in the USA, there were increasing problems for the Ferguson organisation in its arrangements with the Ford company.

Harry Ferguson using a TE-20 to clear weed from a river on his Cotswold estate.

In 1946, the agreement under which the American Ferguson company was reponsible for marketing the Ford 9N tractor was heading for disaster. A new tractor distribution company was formed by Ford to take over the marketing responsibility from Ferguson. The new company, known as Dearborn Motor Corporation, became fully operational in July 1947. At the same time Ford replaced the 9N tractor with a new model, the 8N.

This meant that, from July 1947, Ferguson controlled a large distribution company in the USA which was left without a tractor to sell. One result of this was the famous lawsuit in which Harry Ferguson and his companies claimed more than $300 million from the Ford Motor Company. Another result was the decision by Ferguson to build a factory in Detroit where he could manufacture tractors which his distribution company could market.

While the Detroit factory was being built, TE-20 tractors were imported expensively from Coventry to keep the American market supplied. Then, in October 1948, Harry Ferguson drove the first tractor off the production line at the Detroit factory. This was the first TO-20 tractor, powered by a Continental engine, but sharing the same design as the British tractor.

The TE and TO tractors quickly became one of the outstanding success stories in the history of farm mechanisation, with production rising to a peak in 1951 and 1952 of more than 100 000 tractors a year.

This TE-20 appeared at the 1953 Smithfield Show with Bomford tracks.

They were durable and reliable and prices were kept down in line with Harry Ferguson's strong views on the need for economic stability and the dangers of high inflation.

Because of their reputation for reliability, Ferguson TE-20s were used in 1957 for one of the most spectacular journeys ever made with farm tractors. The journey was an expedition of 1200 miles through some of the harshest conditions on earth to the South Pole. Three Fergusons made the journey with a team of New Zealand drivers under the leadership of Sir Edmund Hillary. The expedition was part of a long-term programme of research and exploration carried out by countries in the British Commonwealth.

There were about a dozen farm tractors in the Antarctic at that time, most of them Fergusons. The first TE-20 arrived there in 1955 with an Australian research team, to be used at their base camp to power generators, move heavy equipment and stores, and to clear sites for buildings. The tractor was kept outside for a year without weather protection, often in appalling conditions, and completed 565 hours' work. The official report by the Australian engineer stated: 'The tractor gave no trouble at all. The terrain and working conditions are tougher than anyone would expect a machine to operate upon with such complete reliability'.

There were few differences between the TE-20s shipped to the Antarctic and the thousands of Ferguson tractors at work elsewhere in the world.

One of several wheel and track arrangements considered for the Antarctic.

One obvious modification was the use of various arrangements of dual front wheels and half tracks to cope with snow and ice. Special insulating material was used on all electrical wiring to withstand the extreme cold and the tractors were equipped with heavy-duty batteries and starter motors.

When the tractors were despatched from the Coventry factory, they were without cabs or any form of weather protection for the driver. Later, shelters were made at the base camp and fitted to some of the tractors. The three Fergusons used for the South Pole journey were also equipped with strengthened track rods, roll bars to protect the drivers if tractors over-turned, and two-way radios to keep the drivers in contact with each other.

Sir Edmund Hillary decided to attempt the journey to the Pole because the reliability of the Fergusons had been so good. The biggest limitation was their pulling capacity, especially in deep snow, which meant they could handle only lightly-laden sledges. To make up for this, the New Zealand team also took a Weasel to help handle the large amount of supplies needed for the journey. The Weasel, a 75-hp tracked vehicle built for the American Army, had been purchased second-hand. During the journey towards the South Pole, the Weasel gave so much mechanical trouble that the New Zealanders were forced to abandon it.

On their journey to the Pole the New Zealand team and their tractors passed through some of the most desolate and difficult terrain in the world. There were areas of deep, soft snow and surfaces which were frozen hard. While crossing a plateau at 10 000 ft, the altitude caused such a serious loss of engine power that the governors had to be reset to 3000 rpm, instead of the usual 2000 rpm, and this increased fuel consumption alarmingly.

In some areas there was danger from crevasses – cracks in the surface which were often wide enough and deep enough to swallow a tractor and driver so completely that rescue would be impossible. Sometimes the crevasses were bridged by fresh snow, leaving a smooth surface masking the danger below. Where crevasses were a problem the tractors were roped together, so that if one broke

Sir Edmund Hillary at the wheel of a TE-20, hauling stores from the supply ship in the Antarctic.

Cabs were fitted to the Fergusons in the Antarctic to keep out the worst of the weather.

Below *One of the three TE-20s which made the journey to the South Pole was returned to England. Here it is being unloaded at London Docks.*

through the surface there was a chance that the others could support it. In the worst areas, the tractors were forced to travel at walking pace behind a man on foot who prodded the surface with a stick to find a safe path.

Sir Edmund Hillary and his team left their base camp on the coast on October 14th and arrived at the South Pole on January 22nd 1958. During the journey, they had carried out some research and also set up a number of supply depots for use by a later expedition.

At the South Pole, the party was greeted by a team of American scientists stationed there for a research project. Sir Edmund arranged for a telegram to be sent to the Ferguson factory at Banner Lane, Coventry. It said: 'Despite quite unsuitable conditions of soft snow and high altitudes our Fergusons performed magnificently and it was their extreme reliability that made our trip to the Pole possible'.

The New Zealanders were flown back from the Pole, leaving the tractors for the American scientists to use. One of the tractors was later flown back and returned to Coventry, where it is preserved in the Massey-Ferguson collection.

MASSEY-HARRIS PONY

The biggest selling tractor produced by Massey-Harris before the merger with Ferguson, was the little one-plough Pony. The tractor went into production at the Woodstock factory in Canada in 1947 and was also produced in France at the Marquette plant from 1951. When the last of the Pony tractors was built in 1961, the production total had reached more than 121 000.

The first production version of the tractor was powered by a four-cylinder petrol engine made by Continental. The gearbox gave three forward ratios and a reverse, with a maximum travel speed of 7 mph, and up to 11 hp available at the drawbar.

Although the Pony was a very basic tractor in the early years of its production, the specification was improved several times during the 14-year production run.

In 1950, the Massey-Harris system of hydraulic implement lift was offered as an option for the Pony. The equipment, known as the Depth-O-Matic system, was the Massey-Harris answer to the Ferguson System and was an external addition to the tractor, not part of the integral design.

Further improvements came in 1957 when the series 820 Pony was announced. This version, built in France, had an improved gearbox with five forward ratios instead of three, together with a diesel engine built in Germany by Hanomag. There were also improvements to the hydraulic lift system, although this still lacked some of the essential features of the Ferguson System. More detailed improvements were added in 1959 when the model 821 went into production.

In North America, where Massey-Harris was most anxious to improve tractor sales, the Pony proved to be too small for most of the farming market. Much of the demand for the tractor came from market gardens, municipal authorities, golf courses and industry.

Against strong competition, especially from the more powerful and more advanced Ford 8N and Ferguson tractors, Pony production in North America reached less than 29 000 and ended in 1954, apart from a batch of 122 manufactured at Woodstock in 1957.

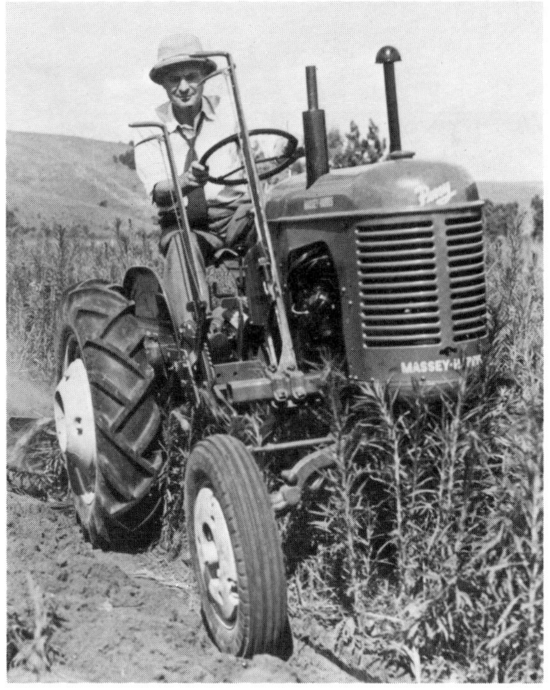

A Canadian-built Pony.

It was the French market which gave the Pony its biggest success story. More than 90 000 tractors were produced from the Marquette factory, most of them going to small farms, which were often making their first break from oxen and horses.

The Pony gave Massey-Harris their first success in the European tractor market and held the company's place in the small tractor sector until the investment in the Ferguson organisation began to produce results.

COUNTY CFT

The County tractor company has an international reputation as a specialist manufacturer of four-wheel drive tractors, but it was the crawler tractor which first brought the company into the agricultural market.

Crawler tractors have never taken a large slice of the market in Great Britain. Their best period followed the ending of World War 2 in 1945, when a demand for improved pulling efficiency caused a significant rise in sales and brought several newcomers into the market. One of the companies which took advantage of the rising interest in tracklayers was County Commercial Cars of Fleet, Hampshire, a company with wartime experience of building off-the-road vehicles for the British Army.

County's first farm tractor was completed in 1948. It was built around the popular Ford E27N engine, starting a connection between County and Ford which still survives. The tractor was called the County Full Track, later shortened to CFT.

The four-cylinder Ford engine, developing 36 hp on petrol and 30 hp on paraffin, was later replaced by the Perkins P6 diesel engine. The Fordson gearbox was also used in the County, giving the tractor a maximum forward speed of 3.38 mph in third gear. The final drive consisted of a spiral bevel to the countershaft and a straight spur-gear drive to the rear axle.

The standard track width was 12 in, with alternative widths available. There were five track rollers on each side, all with replaceable steel tyres. The ground pressure under the standard tracks was 4.5 psi and the claimed maximum drawbar pull was 6000 lb.

A feature of County tracklayer design was the height of the top of the track above the ground. This made it more difficult to climb into the driving seat, but it also gave some practical advantages. It was claimed that the large front wheel and sprocket gave a better performance over uneven surfaces and also reduced the angle of movement in the track links helping to minimise maintenance costs.

The high profile tracks also helped to give County tractors a distinctive appearance which survived until wheeled tractors took over. The successful CFT tractor remained in production until 1952, when the model Z tracklayer was introduced, based

County CFT.

on the New Fordson major. Other tracklayers from County included the Ploughman of 1957 and various export and industrial models.

Four-wheel drive, offering much of the pulling efficiency of a tracklayer but with better versatility, began to affect demand for crawler models. County was one of the companies helping to establish the four-wheel drive concept in Great Britain when the County Fourdrive wheeled tractor arrived on the market in 1954. County's interest in four-wheel drive developed rapidly and production of tracklayers ended in 1964.

County Full Track at work on a hill land improvement scheme in Devon.

14 (Above) *Farmall BMD.*

15 (Below) *A Field Marshall Series 1.*

16 (Above) *Fordson E27N Major.*

17 (Below) *Ferguson TE-20 built in 1950.*

18 (Above) *A Fordson E27N Major with half-track conversion.*

19 (Below) *The only Hurlimann D-100 exported to Britain is now preserved at Mr Philip Jenkinson's museum, Shebbear, Devon.*

20 Massey-Harris 744PD, one of the first batch built at Manchester.

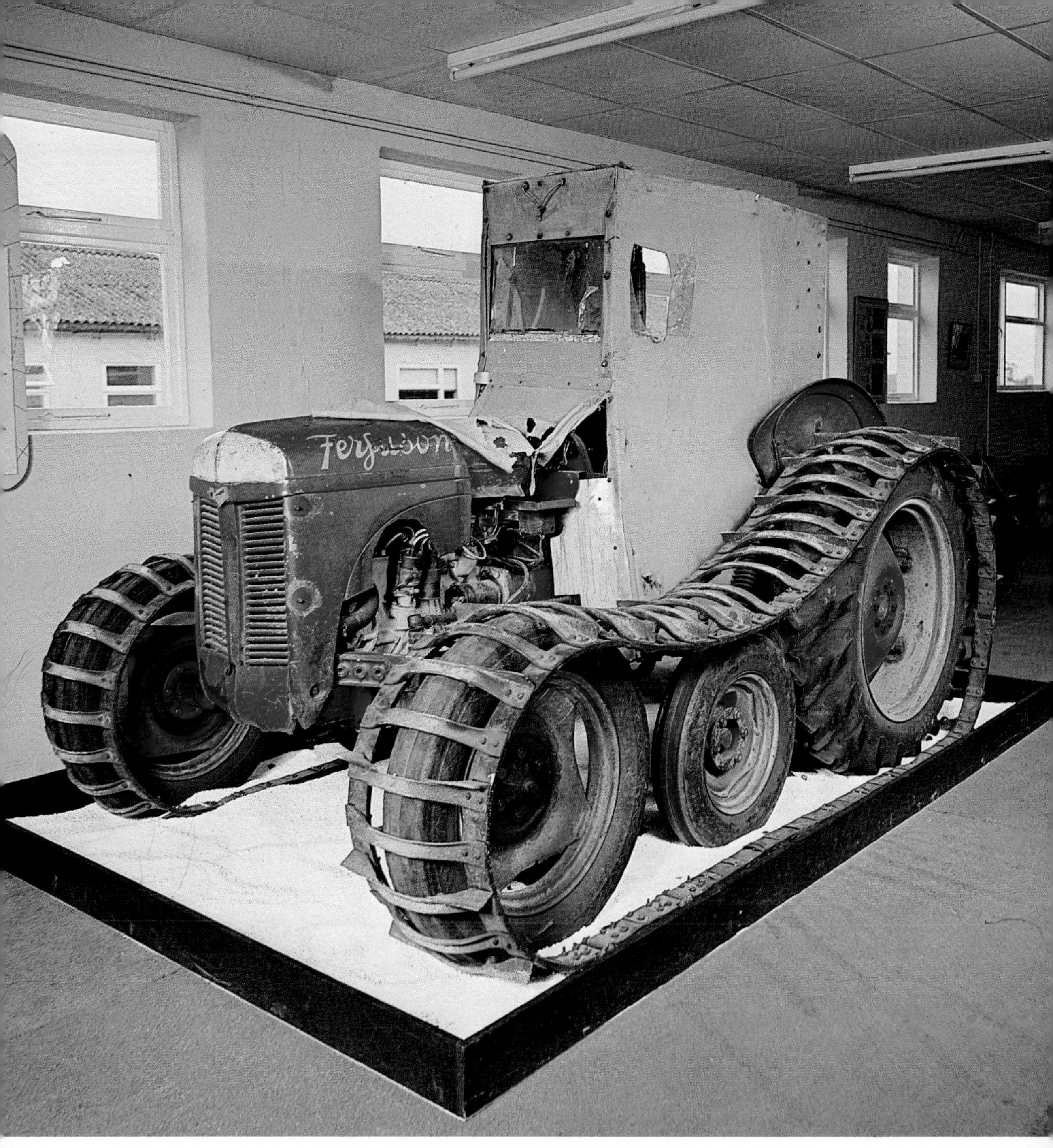

21 Back from the South Pole, this Ferguson tractor is now preserved in the Massey-Ferguson museum at the company's training school at Stoneleigh, Warwickshire.

22 (Above) *Massey-Harris Pony built in France.*

23 (Below) *Turner tractor built in 1948.*

24 (Above) *BMB President sold new in 1956.*

25 (Below) *A Marshall MP6 built in 1954.*

26 (Above) *Allis-Chalmers D270*.

27 (Below) *Allis-Chalmers D272*.

NUFFIELD UNIVERSAL SERIES

William Morris made a reputation and a fortune by building up one of the most successful companies in the British motor industry. Later he established a second reputation as a philanthropist, when he gave much of his wealth to charity.

He controlled a substantial share of the British market with his Morris, MG, Riley and Wolseley cars, and Morris Commercial became a leading make in the truck market. Morris, who later became Lord Nuffield, had taken an early interest in the tractor market. In 1926, his company had produced a prototype crawler tractor based on a small tank developed for the British Army. This project had been abandoned, but another attempt to move into the tractor market was started as World War 2 was approaching its final stages.

Work on a prototype for the new tractor was well underway by the end of 1945 and the first completed machine was taken to Lincolnshire for field-testing in the following year. By the end of 1946, there were 12 prototypes at work. The test results were encouraging and space was made available at the Wolseley car factory in Birmingham to put the new tractor into volume production.

The tractor was announced at the 1948 Smithfield Show, in London, as the Nuffield Universal. It was available as the M4 in conventional four-wheel design and as the M3 with tricycle wheel arrangement for rowcrop work.

The new Nuffield was equipped with a four-cylinder, side-valve engine, starting on petrol and running on paraffin. This was based on the Morris Commercial-type ETA engine which had been the power unit for thousands of British Army trucks during the War. The engine developed 42 hp at the maximum 2000-rpm governed speed. An alternative power unit was offered from 1950 to meet the growing demand for a diesel engine. The diesel unit chosen was a Perkins P4, but this was replaced in 1954 by a new 3.4-litre diesel made by British Motor Corporation, the new company which had been formed by an amalgamation of the Morris organisation with the Austin Motor Company.

Another development of the Universal series came in 1957 when a smaller tractor with similar styling was added to the range, known as the Universal 3, available with a diesel engine only. This was also a BMC diesel, 2.5-litre capacity in three cylinders. At the same time, the larger model became known as the Universal 4 and some detail improvements were made to the hydraulic system.

The Universal was announced as 'Britain's most

Nuffield Universal with BMC diesel engine.

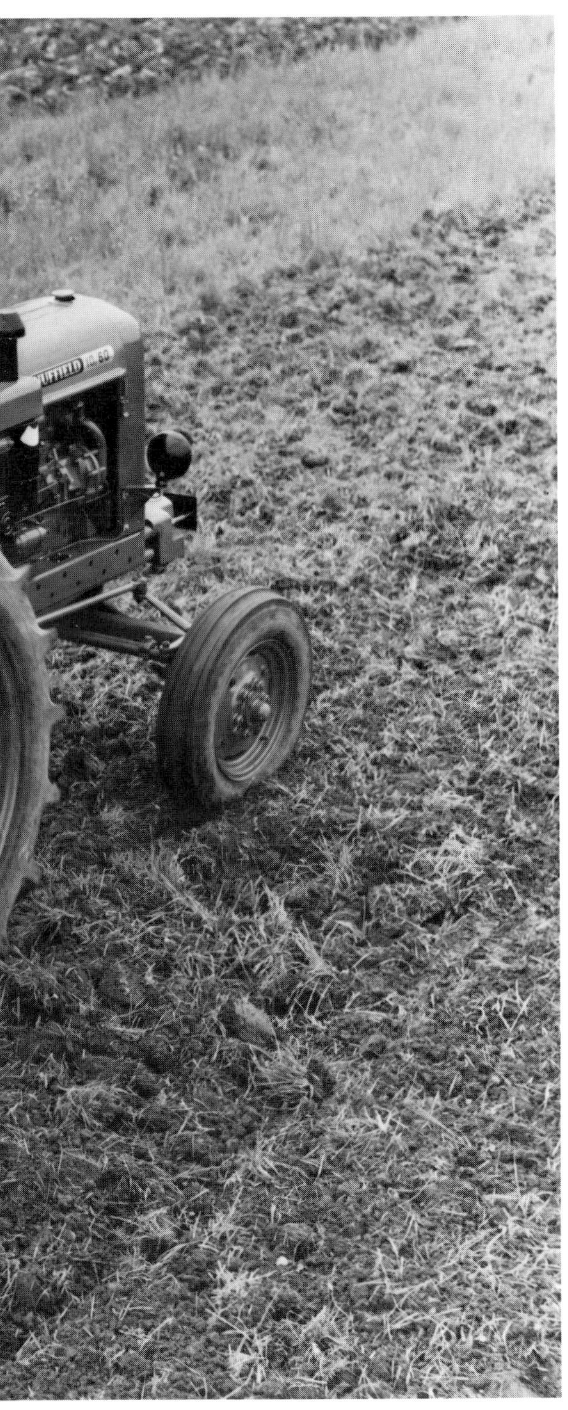

powerful tractor' in 1961 when the Universal 4/60 was announced. This was rated at 60 hp, the increase being achieved by boring out the 3.4-litre BMC engine to 3.8 litres. At the same time, the transmission was strengthened to take the extra torque of the more powerful engine, and there were further improvements to the hydraulic system and three-point linkage.

Production of the Nuffield tractor range had been transferred from Birmingham to the Morris Motors factory at Cowley, Oxford. As part of the BMC reorganisation, production was again moved, in 1963, this time to the new Bathgate factory in Scotland. This made the Universal the first tractor to be produced in Scotland on a large scale, apart from the ill-fated Glasgow.

By now the Universal series was starting to look dated compared with some of its rivals and sales were beginning to suffer. As there was no immediate prospect of a completely new model, the decision was taken instead to give the ageing Universal a thorough facelift. The result was announced in 1964 when the Nuffield 10–60, and the smaller version – the 10–42 – were both put into production.

The most significant improvement was the addition of a hi-lo ratio gearbox to give a total of ten forward ratios and two reverse. This was a development which was already available on many other makes and it helped to bring the Universal into a more competitive position on the market.

Some of the other improvements were also important, including disc steering brakes, an improved drawbar design and a new hydraulic control unit incorporating depth control.

This was the last version of the Nuffield Universal series and it survived until 1967, when a completely new range of Nuffield tractors was announced at the Royal Show. While the new Nuffields were becoming established on the market, the parent company was once more involved in a merger. This time the result was the British Leyland Motor Corporation, which soon finished off the production of the Nuffield tractor, replacing it with the more stylish Leyland. Leyland Tractors was sold in 1981 to join Marshall as part of the Nickerson organisation.

The 10–60, most powerful of the Nuffield Universals.

TURNER 'YEOMAN OF ENGLAND'

The Turner Manufacturing Company of Wolverhampton announced their first tractor in 1949 at the Royal Show. Some time after the initial launch of the tractor, it was give the name 'Yeoman of England', which appeared on a distinctive brass plate just behind the radiator.

Another distinctive feature of the Yeoman was its power unit. This was a four-cylinder diesel with the cylinders in twin banks of two in a 68° vee-arrangement. The engine bulged out on both sides of the tractor as an easily recognised feature which helped to create an impression of power.

The engine was designed by Freeman Sanders, who was one of the outstanding British diesel engineers. Sanders had introduced an improved combustion chamber in 1935 while working for Fowler of Leeds and this had proved to be a major step forward, giving much better combustion with smoother running and cleaner exhaust. After contributing a great deal to Fowler's success with diesel engines, Freeman Sanders had left the company in 1941 to set up his own small engineering design organisation in Penzance, Cornwall. His vee-4 diesel engine was manufactured by Turner first for industrial and marine use and later as the power unit for their new tractor.

With a cubic capacity of 3271 cc, the Yeoman engine was rather smaller than it appeared. It was rated at 34 hp at 1500 rpm, with a maximum output of 40 hp at 1725 rpm. The maximum drawbar pull was claimed to be 4500 lb at 1.9 mph.

One of the thoughtful features of the Turner design was offsetting the driver's seat and steering wheel. This allowed a less obstructed line of forward vision to the right of the engine.

In spite of what was probably a good engine and an impressive appearance, demand for the tractor proved to be disappointing. There appear to have been several reasons for this, including price and specification and some mechanical problems.

The Yeoman was not a cheap tractor and the basic model in 1950 cost £669, which was significantly more than its obvious competitors in the British market. The specification of the standard Yeoman model left a good deal to be added as optional extras. Rubber tyres, hydraulics, three-point linkage, power take-off, belt pulley and even the differential lock were all additions to the basic model and to the basic price.

Most of the mechanical problems appeared in the early production versions of the tractor and were dealt with quickly. These included a weakness in the drive to the pulley and a tendency to overheat because of insufficient radiator capacity. There was also a weakness in the main transmission

which appears to have been more persistent and troublesome.

In spite of a face-lift in 1951, when some detail modifications were introduced, the Turner tractor never really achieved the commercial success it probably deserved and production ended after about 8 years.

Turner 'Yeoman of England' tractor.

OTA

One of the most popular of the post-War small tractors was the OTA which arrived on the market in 1949. The name 'OTA' was derived from the name of the company which manufactured it, Oaktree Appliances of Fillongley, Coventry.

The original version of the tractor weighed only 13 cwt and was designed to take the immensely popular Ford 10 engine. The industrial version of the engine was used, and a paraffin version became available from 1950. One advantage of the three-wheel design and small size of the tractor was a high degree of manouevrability and the tractor was claimed to have a 12-ft turning circle.

Some of the small tractors on the British market at that time were offered with very basic equipment, but the OTA had an unusually comprehensive specification. There was a double-reduction gearbox, with six forward ratios and two reverse, giving a speed range of up to 13.4 mph. Also part of the standard equipment was an hydraulically operated implement lift, included in the basic list price of £259.

Although the OTA was reasonably successful in a highly competitive market, it became increasingly obvious that there was a larger market for a four-wheel tractor with more power and with more real work potential. To meet this demand, and to compete more effectively with the Ferguson, Oaktree Appliances introduced the Monarch in 1951.

A 17-hp Ford engine, available in petrol or petrol/paraffin versions, provided the extra power for the Monarch. Like its predecessor, the Monarch was equipped with six forward gears and with similar 6-in × 22-in rear wheels. But the hydraulic lift had become an optional extra, as was the four-speed pulley and the 550-rpm power take-off.

The Monarch was no match in price or performance against the Ferguson, Fordson and other popular farm tractors and Oaktree Appliances withdrew from the market in about 1955. There was a possibility that the Monarch might have been taken over by the Singer car company, which wanted to diversify into the tractor market. This project fell through because Singer was experiencing problems at that time and became part of the Rootes Group in 1956.

The little OTA tractor from Coventry.

BMB PRESIDENT

Although the President remained in production only 7 years or so, it was more successful than many of the small economy tractors which were appearing on the British market at that time.

The new tractor was announced at the 1950 Royal Show. It was manufactured by the Brockhouse Engineering Company of Southport, Lancashire. The President was the first tractor from Brockhouse, although the company already produced a range of hand-controlled tillers.

Brockhouse designed their new tractor as a low-priced, basic power unit for small farms and market gardens. They chose a Morris 8/10 engine, a four-cylinder unit of 919-cc capacity. The engine was basically the familiar Morris car engine, developing 16 hp at 2500 rpm on petrol or 14 hp on paraffin. The tractor weighed 1800 lb and produced 2500-lb maximum pull at the drawbar.

The tractor was equipped with three forward gear ratios and a reverse, with a top speed of 8.3 mph. In line with the low price approach, the basic model did not include a power take-off or three-point linkage, although these were available as extras.

In 1955, the list price of the basic tractor was £229.10s (£229.50), which looked reasonably competitive. But demand was failing to develop as Brockhouse had planned and the commercial future for the tractor was becoming doubtful.

Two years later an attempt was made to relaunch the President. The Brockhouse company apparently transferred their interest in the tractor and the new version was launched at the 1957 Smithfield Show as the Stokold President, backed by the H.J. Stockton Company of London and shown at Smithfield by the Owen Organisation.

The Stokold tractor was based on the old Brockhouse design. The main change, underneath the more stylish exterior, was a two-cylinder Petter air-cooled diesel engine. The engine, started by hand, developed 13.8 hp.

Twenty years after the Stokold tractor was introduced, small diesel-powered tractors from Japan had begun to sell in large numbers in the USA and Europe. But in 1957 there was little sign of success for the Stokold President and the revival attempt failed to develop.

This attempt to revive the BMB President as the Stokold, appeared at the 1957 Smithfield Show, but it is unlikely that any were sold.

FIAT 55

By European standards, the Fiat 55 crawler was a large tractor when it arrived on the market in 1950. For Italian conditions, where there is a tradition of deep ploughing on heavy soils, the new Fiat met a predictable demand. The export prospects were less obvious, especially to countries such as Great Britain where crawler tractors have rarely achieved more than a tiny percentage of total sales.

In fact, the Fiat 55 was a success at home and abroad and, from 1951, when it first arrived on the British market, the 55 helped to re-establish Fiat tractors after an absence of nearly 30 years.

The tractor was available in two versions, the 55 and the 55L. Both were basically similar, but the L model had lower gearing to give extra pulling power at the drawbar, and also wider tracks for reduced ground pressure.

A four-cylinder diesel engine of 6.5-litre capacity developed 55 hp at the maximum 1400 rpm, driving through a gearbox with five forward speeds and a reverse. The transmission included a central pair of bevel gears, with the final drive through spur gears.

Fiat offered the unusual choice of steering by means of levers or with a steering wheel. Another unusual feature of the design was the electrical system which had no battery. To provide current for the lighting system, the 55 was equipped with a 6-volt, 90-watt dynamo and, instead of an electric starter motor, there was a small donkey engine.

A disadvantage of the donkey engine was that it ran on petrol, which meant having two types of fuel available. The small engine was a horizontally-opposed twin-cylinder unit made by Fiat and developing 10 hp. The donkey engine was started by pulling a rope and was equipped with its own magneto. It engaged against a ring gear on the flywheel of the main engine.

The donkey engine was water-cooled and the cooling systems of the main diesel engine and the small petrol engine were linked together. This meant that the smaller engine provided warm water to help raise the temperature of the diesel for easier starting in cold weather. Once the big engine had started, the drive from the donkey engine to the flywheel disengaged automatically.

Fiat aimed their big crawler tractors at large arable farms in Europe, as well as at forestry and industrial concerns. Their customers required pulling power at the drawbar and the 55 and 55L tractors performed well. The maximum drawbar pull of the 55L was 12 125 lb.

Fiat 55 crawler being demonstrated in Berkshire.

FOWLER CHALLENGER SERIES

The Challenger project was an ambitious and complicated plan to put Fowler firmly back into the tractor market and to provide a product range to keep the big factory at Hunslet, Leeds, in business.

When the Challenger tractors were under development, the Fowler organisation was already part of the Thomas W. Ward group, following the merger in 1947 with the Ward subsidiary, Marshall of Gainsborough. In an effort to rationalise the Fowler and Marshall product ranges, the Fowler factory was given responsibility for building crawler tractors. This was a product in which the company had considerable experience.

Four models were included in the Challenger range and these were all planned at the same time. The first to be announced was the Mark 3 Challenger, which was given priority because it

Fowler Challenger Mark 1.

was expected to have the biggest sales potential.

The Mark 3 was launched at Smithfield Show in 1950 with a 95-hp Meadows diesel engine. A Leyland AU600/9 diesel, de-rated to 95 hp, was available as an alternative. The power was delivered through a 16-in single-plate clutch and a gearbox with six forward ratios as standard or twelve as an option. The Meadows version weighed 23 600 lb, and the list price when production started in 1951 was £3525.

Smallest of the Challengers was the Mark 1, which was shown at Smithfield in 1951 and was the model with the greatest agricultural potential. The engine was an unusual twin-cylinder, two-stroke diesel made by Marshall and known as the ED5. It was a 'scavenged' design, with a Rootes blower and no valves.

The Challenger Mark 2 with AU350 engine.

It was claimed that the ED5 engine combined the mechanical simplicity of a ported two-stroke with the smoothness and quietness of a four-cylinder, four-stroke engine. The Marshall engine was provided with a small Coventry Victor petrol engine for starting.

An unusual feature of the Mark 1 design, resulting from its role as a civil engineer's tractor as well as a farm crawler, was the provision of four power take-offs. There was a p-t-o fitted as an extension of the crankshaft at the front of the tractor, and intended mainly for powering front-mounted hydraulic equipment. A side p-t-o was fitted as an extension from a gearbox shaft, giving a speed of 686 rpm at standard engine speed. Number three was a conventional p-t-o shaft at the rear of the tractor, offering a choice of 535 or 1250 rpm. The fourth power take-off was described in the sales leaflet as an 'emergency' fitting and was driven by a chain from the fan pulley.

Fowler announced the Mark 2 Challenger in 1950, but production was delayed for more than 12 months. A Leyland AU350 six-cylinder diesel engine, rated at 65 hp, was used as the power unit. The Mark 2 appears to have been dropped from the Challenger range within about 3 years.

At the top of the Challenger range was the Mark 4, weighing 14.5 tons, with a 15.9-litre Meadows engine rated at 150 hp. Production started in 1953 and most of the Mark 4 tractors were sold for export, particularly for land clearance and civil engineering work.

In an effort to find new markets for the Challenger tractors, Fowler introduced a special version of the Mark 3 which was designed for launching lifeboats and hauling them back from the water again. The modifications, which were available from 1953, included sufficient waterproofing to allow the tractor to go into shallow water, steering controlled by a ship's-type wheel instead of by levers, and a winch as a standard fitting. The winch was used for pulling the heavier lifeboats, some weighing as much as 20 ton, up the slope from the water to the boathouse.

Although the Challenger tractors were fairly successful in the highly competitive world market for large crawler tractors, they provided insufficient profit or production volume to keep the Fowler factory open. The factory eventually closed in 1974 and the Fowler name disappeared from the tractor market.

Fowler Challenger Mark 3.

CHALLENGER

Challenger Mark 4 with 150-hp Meadows diesel engine.

COUNTY FOURDRIVE

County Commercial Cars made their first break from crawler tractor production in 1954 when they announced the Fourdrive, a large-wheel tractor which was the start of the company's development as a specialist in four-wheel drive.

Now there is plenty of evidence to show that four-wheel drive can provide extra pulling efficiency compared with conventional two-wheel drive. The four-wheel drive advantage is greatest with four large wheels, as preferred by County and some other British manufacturers, rather than with smaller front wheels as generally used by European tractor firms.

In 1954, there was little factual evidence available on tractive efficiency and the four-wheel drive market in Great Britain had not begun to develop significantly. County designed their Fourdrive tractor for export, particularly for the sugar-cane areas where large quantities of material have to be transported.

The Fourdrive was based on the New Fordson Major engine and also inherited ideas from County's growing experience with tracklayers. The most unusual feature of the Fourdrive was that it was skid-steered like a crawler, with steering levers instead of a steering wheel.

With a more conventional four-wheel drive tractor, the designer has to combine both a steering and a driving system to the same wheels, and articulated tractors have been developed partly to avoid this problem. The skid-steer Fourdrive also avoided the problem, allowing a simpler, stronger, cheaper axle design.

The limitations of the skid-steering system ruled

County Fourdrive.

the Fourdrive out as a general-purpose tractor for European conditions, but it was successful for the sugar estates for which it had been designed and also for certain other situations, such as forestry, where heavy loads have to be moved, sometimes in difficult ground conditions. County claimed a drawbar pull of 7750 lb, which was considerably greater than the pulling power claimed for the two-wheel-drive Fordson tractor with the same engine.

The four-wheel-drive tractor used the Fordson gearbox, which allowed travel speeds up to 11 mph on the road. County's sales of crawler tractors for sugar-cane work had been limited by slow travel speeds, restricted use on roads and by the high cost of track maintenance caused by the ridges and ditches of cane-growing areas.

Although the market for the Fourdrive was restricted, experience with the tractor encouraged County to take a closer look at four-wheel drive. The Super Four and Super Six tractors, both with four big driving wheels, but with front-wheel steering, were on the market from 1961, while production of crawlers and the Fourdrive came to an end in 1964.

MARSHALL MP6

The MP6 was an ambitious attempt by Marshall of Gainsborough to move into the top end of the tractor market with a really powerful new model. The company was already well established with the familiar single-cylinder diesel tractors which were particularly popular for stationary work with a belt or winch. The MP6 was designed primarily for drawbar work, and was among the most powerful wheeled tractors in the world when it was announced in 1954.

Although the MP6 was shown for the first time

Most of the MP6 tractors were exported, like this one hauling sugar cane.

The Marshall MP4 was built and tested with a Meadows four-cylinder engine, but this model never went into production.

at the 1954 Smithfield Show in London, the first tractors were not available for delivery until two years later.

The new tractor caused considerable interest because its size and power were so impressive by 1954 standards. Marshall used a six-cylinder Leyland diesel engine for the MP6. This was the U/E 350 engine giving the tractor almost 7000 lb of drawbar pull. The overall weight of the MP6 was 12 600 lb.

When the MP6 was tested by the NIAE at Silsoe, the fuel consumption averaged 3.51 gallons an hour with the engine developing 69 hp at 1700 rpm.

In spite of the interest the new tractor caused, the sales demand was disappointing. Marshall expected a good export market for the tractor because of its obvious advantages on large acreages. To some extent this proved correct and tractors were sent to Australia, Spain and other markets, including sugar estates in the West Indies. British farmers were not yet ready for such a powerful and expensive tractor and sales on the home market remained small.

The limited export successes and a new long-wheelbase version developed for forestry, failed to save the MP6 and the decision was taken to end the production run in 1960. This was the last wheeled tractor produced by Marshall, who had built their first wheeled tractors in 1907. After 1960, the company concentrated on crawler tractors, the last British manufacturer to specialise in this sector of the market.

NIAE HYDROSTATIC TRACTOR

The use of hydraulic systems for transmitting power has been available on farm tractors for many years. The Ferguson System of three-point linkage and automatic draft control is based on the use of hydraulics, and power-assisted steering and the front-end loader are also hydraulic systems.

Hydraulic systems can also be used to provide hydrostatic transmission to turn the wheels of tractors and other vehicles. This is an idea which has attracted engineers since the beginning of the century, but it has only recently reached the production line.

Much of the development work which helped to bring hydrostatic transmissions on to the farm was carried out by the British National Institute of Agricultural Engineering at Silsoe, Bedfordshire. An experimental tractor with hydrostatic transmission was demonstrated by the NIAE in 1954, after a 2-year development programme. This was based on a Fordson Major tractor with a diesel engine.

In a hydrostatic transmission, power from the engine is used to drive a pump which forces oil, through a series of pipes, to a hydraulic motor. The oil, under pressure from the pump, moves pistons in the motor and turns the driving wheels of the tractor. This was basically the system used for the NIAE prototype tractor, but variations have been introduced on more recent hydrostatic tractors.

The whole of the transmission of the Fordson was removed when the prototype was being built. The pump was mounted on the blank nearside of the cylinder block, driven by a chain from the crankshaft. The NIAE used two motors, each a five-cylinder radial design built to fit into the driving wheels of the tractor.

When the NIAE tractor first showed its paces in public, it was welcomed as a significant breakthrough in tractor design. There were many who expected hydrostatic drives to oust the old-fashioned gear transmission from popularity.

The fact that there is no mechanical link between the engine and back wheels means that a hydrostatic transmission is unlikely to be damaged by overloading or inexperienced operators. The fluid drive gives infinitely variable speed between zero and the maximum, without adjusting the

The NIAE Hydrostatic Tractor built around Fordson components.

throttle or breaking the drive to change gear. The lack of a clutch and gear levers simplifies the job of the driver and, finally, the type of drive used by the NIAE has a particular advantage for powering tractor wheels which also have to be steered, and is of interest for four-wheel drive.

These advantages are now on the market in the form of hydrostatic drives on two- and four-wheel-drive tractors, combine harvesters and other specialised farm and garden equipment.

The fact that commercial development has made slower progress than was once expected is partly because of the slightly greater purchase and operating costs of hydrostatic transmission. It has also been affected by the important progress which has been made in developing the conventional gear drives to make gearboxes easier to use, smoother and more efficient.

The NIAE tractor ploughing with hydrostatic drive.

ALLIS-CHALMERS D270 AND 272

The British tractor industry expanded rapidly after the end of World War 1 as new factories were opened to meet the rapidly growing demand for tractors for export as well as for British agriculture. American companies were part of the expansion, including Allis-Chalmers, which opened an assembly plant for the Model B tractor near Southampton in 1948.

The Model B was already a dated design by 1948 and it was replaced in 1955 by the D270. This tractor was produced at Essendine, Lincolnshire, where the company had moved in 1953 to a larger factory close to the main arable farming areas of England.

Although the styling of the D270 looked new, much of the design was based firmly on the Model B. This included the high clearance of the D270, which allowed the use of mid-mounted implements for rowcrop work, and the distinctive narrow 'waist' for good visibility from the driving seat.

There was a choice of three engines for the D270, all of them inherited from the later production version of the model B. The engine options included the petrol and paraffin versions of the Allis-Chalmers four-cylinder, overhead-valve engine with $3\frac{1}{3}$-in bore and $3\frac{1}{2}$-in stroke. The petrol version developed 27 hp at 1650 rpm and the output from the paraffin engine was 22 hp at the same speed. For their diesel engine, which cost an extra £145.10s (£145.50) in 1955, Allis-Chalmers used the Perkins P3/143, later replaced by the P3/144 which developed 31 hp at 1900 rpm.

One new feature which had not already appeared on the B tractor was the live power take-off. This meant that the forward motion of the tractor

Allis-Chalmers publicity picture to show the use of mid-mounted equipment on the D272.

could be halted, by means of an auxiliary hand-operated clutch in the right hand half-shaft, without affecting the power to the p-t-o, pulley or hydraulics. The drive to the p-t-o was controlled by a separate hand-lever to the right of the driver's seat, operating a dog clutch in the upper shaft of the gearbox. The live power take-off was especially useful for working with the Rotobaler, the round baler which scored such a big commercial success for Allis-Chalmers long before the modern generation of big round balers. The D270 gearbox was also an improvement, with four forward ratios instead of the three which were standard in the model B.

Allis-Chalmers announced the D272 in 1959, still with the same choice of engines, except that the petrol and paraffin versions could be operated at up to 1900 rpm to increase their power output to 30 and 26 hp respectively. There were other improvements, notably to the hydraulic system, but the attractive styling of the D270 was retained.

A new model, the ED40, arrived on the market in 1960, but sales of this tractor, as well as of the D272, were disappointing. Allis-Chalmers stopped producing agricultural tractors in Great Britain in 1968, although they continued for several years to produce farm machinery. The Essendine factory is now part of the Fiat-Allis organisation.

DAVID BROWN 2D

One of the new tractors announced at the 1955 Smithfield Show in London was the 2D on the David Brown stand. The extremely successful Cropmaster had reached the end of its production run in 1953 and the 2D was one of a number of new models introduced by David Brown to widen the company's product range and fill the Meltham factory.

The 2D was a tool carrier, a specialised type of tractor designed for the accurate, inter-row work in market gardens and rowcrop farming. It is a relatively small sector of the total tractor market which is usually supplied by specialist companies operating on a small scale. When the 2D was withdrawn from the market in 1961, only 2008 had been produced, which was a low volume for a major tractor manufacturer but represented a large share of the available market.

Above and left. *An air pump on the 2D provides pressure to raise and lower implements, with the main frame acting as the air reservoir.*

Important design features of a tool carrier include good visibility and versatility. To provide a good forward and downward view from the tractor seat, the engine of the 2D was positioned behind the driver. This helped to give the driver good control over the operation of equipment worked from a mid-mounted toolbar.

A wide range of hoes, seeders and chemical applicators was available for the 2D and a second toolbar could be fitted to the rear of the tractor to work in conjunction with the mid-mounted equipment. This meant that several jobs could be completed in one operation.

The mid- and rear-mounted toolbars were raised out of work by means of compressed air, with two lift-cylinders for each toolbar. The mid-mounted toolbar could be operated independently of the rear unit if necessary. The reservoir for the compressed air system was the tubular metal framework of the tractor, which appears to have been a satisfactory and reliable arrangement.

An unusual feature of the David Brown design was the use of a diesel engine, at a time when tool carriers were usually petrol or paraffin-powered. The diesel engine was made by David Brown and was an air-cooled twin-cylinder unit developing 14 hp and driving through a four-speed gearbox. A rear lift-linkage and a power take-off could be specified as optional extras.

Although the 2D production run ended in 1961, many are still in regular use. Some tractor dealers in the main market garden areas of Britain meet a steady demand for reconditioned 2D tractors.

David Brown 2D tractor.

BROWN

JOHN DEERE 820

John Deere '20' series tractors were announced in 1956 in a range of six different basic models. At the bottom of the range was the little one-plough 320. The 820 was then the big tractor in the product line, meeting a growing demand for more power on the larger arable farms of North America.

These were the first John Deere tractors with a two-tone colour scheme. This was the most important difference in appearance, as the '20' series tractors retained much of the styling of the previous range. Technical differences included improved hydraulics, plus very much more comfort and convenience for the driver. Compared with previous John Deere tractors, the '20' range provided a better instrument layout, more foot-room and a seat which was fully adjustable and sprung.

Only the 820 tractor was diesel-powered, but even this model still used the traditional John Deere engine layout of two horizontal cylinders. The company had built their first diesel tractor in 1949. This was the famous model R, and the 820 was the direct descendant of that tractor. The engine in the R had developed 43.3 hp at the belt, but the 820 engine developed 64.3 hp, making it the most powerful farm tractor in the company's history at that time.

Capacity of the diesel engine was 471.5 in^3 with $6\frac{1}{8}$-in bore and 8-in stroke. The rated engine speed was only 1125 rpm. There was a small petrol engine for starting the main diesel engine. This was a vee-4 design which had been introduced on the R.

As the 820 was designed mainly for drawbar work, the power take-off was still considered an optional item. The hydraulic lift, with 1430-psi operating pressure and 13 gallons per minute delivery, was standard.

Demand for power, especially in the important sales areas for John Deere in the mid-west, was beginning to develop in the 1950s and the 820 was designed for that market. Although the 820 remained in production for less than 3 years, its power was stepped up during that time to keep up with the market trend. The tractor weighed more than 4 tons and was claimed to pull six 14-in furrows or a 21-ft set of discs.

The short commercial life of the 820 ended in 1958 when the '20' series was replaced by the '30s'.

John Deere 820.

DOE TRIPLE-D AND 130

During the 1950s, most tractor manufacturers in Great Britain and Europe were concentrating their efforts on the mass market and little effort was made to meet a demand for more power and the extra pulling efficiency of four-wheel drive. A wide choice of crawler tractors was available but for some farms, and on some soils, these were not ideal.

One of the farmers who found the selection of tractors inadequate for his needs was Mr George Pryor of Navestock, Essex. He farmed a large arable acreage, much of it on heavy Essex clay, and he wanted the power and pulling efficiency to cope with cultivation in conditions which were often difficult.

After taking a careful look at the tractors available in the mid-1950s, Mr Pryor decided to design one of his own. The result, built in his farm workshops, was an articulated tractor made up from two Fordson Majors linked together. The front of one tractor was joined to the back of the other, with the front wheels of both units removed. The driver sat on the seat of the rear tractor with the power of the two engines at his finger-tips. The complete unit was steered by turning the front section, using a hydraulic ram controlled by the steering wheel of the rear section.

The complete tractor unit had some disadvantages. There were two complete engines to check over and to service, two fuel tanks to fill and, on the prototype, the front engine could not be started from the rear driving position.

Much more important were the advantages. At a time when 50 hp meant a large tractor and four-wheel drive was almost unobtainable, Mr Pryor's home-made tractor could out-perform other wheeled tractors on the British market. In spite of its enormous length, the articulated tractor proved to be highly manouevrable.

Among the people who came to see the tractor working was Mr Ernest Doe, the owner of a highly successful farm equipment retailing company with the Ford tractor agency which dealt with most of Essex. Mr Doe was impressed by the performance of the articulated tractor and realised that other heavy-land farmers might be keen to buy a similar unit. An agreement was signed to allow Ernest Doe

The articulated Doe Triple-D tractor with two Fordson Power Major engines.

DOE'S DUAL POWER

& Sons to produce tractors based on Mr Pryor's design.

Production started in 1957. The articulated tractor was called the Triple-D, standing for the Doe Dual Drive, and it was sold locally at first, meeting a brisk demand from farmers in Essex and adjoining counties. Pleased with the success of the new tractor, the Doe organisation began selling it nationally. There were also some export orders and Triple-D tractors were shipped to Australia, West Africa and Scandinavia.

In 1960, the Doe tractor won a Silver Medal at the Royal Agricultural Society of England's Royal Show, one of only two silver awards made that year.

When the Fordson Major series was replaced by the Ford 4000 and 5000 models, a new Doe tractor was introduced. This was the Doe 130, consisting of two Ford 5000 engines providing a total of almost 130 hp.

Other tractor companies were soon offering additional power and four-wheel drive, providing the Doe company with increasing competition. Price was becoming a problem, as the twin engine design was an expensive arrangement which more conventional designs avoided.

More than 300 articulated tractors were built by Doe and many of these are still working. The Doe 130 illustrated, new in 1965, was built when sales were falling.

The final form of the Doe tractor, the 130.

READING UNIVERSITY DRIVERLESS TRACTOR

Several systems have been used experimentally to develop tractors which can operate without a driver, and most of these projects have made use of radio signals to actuate the tractor controls, a method widely used to operate model aircraft.

A British research team from the Farm Mechanisation Department of Reading University chose a different method when they were planning an automatic guidance project. With radio control, they argued, the man on the tractor seat was replaced by a radio-operator and there was an advantage only when one radio could be used to control several tractors.

Work on the Reading project continued for many years, but preliminary details were released when the system was demonstrated publicly in 1958. The control system was based on a network of energised wires which were buried below the ground surface along routes frequently used by the tractor on one of the University farms.

An International Harvester B250 tractor was modified to follow the wires. At the front of the tractor were two search coils which detected the current in the buried wires. When the centre of the tractor was exactly over the wire, the two coils were in balance. If the tractor was not correctly positioned over the wire the coils went out of balance and this happened if the steering deviated or if the route taken by the wire turned.

The search coils were in circuit with a balance relay which controlled a solenoid. The solenoid was linked to hydraulic valves which governed the oil supply to a double-acting ram in the tractor steering linkage, a form of power steering.

It was claimed that the Reading University project was the first attempt to adapt a farm tractor to follow a wire guidance system and the experiment was technically successful. The tractor could be made to follow a complex pattern of routes with an acceptable level of reliability and the wire could also be used to transmit other messages to the tractor as well as controlling its steering. Signals in the guide wire could stop and restart the tractor, operate the hydraulic lift mechanism and the power take-off and also control the forward speed.

Although the demonstrations showed that the

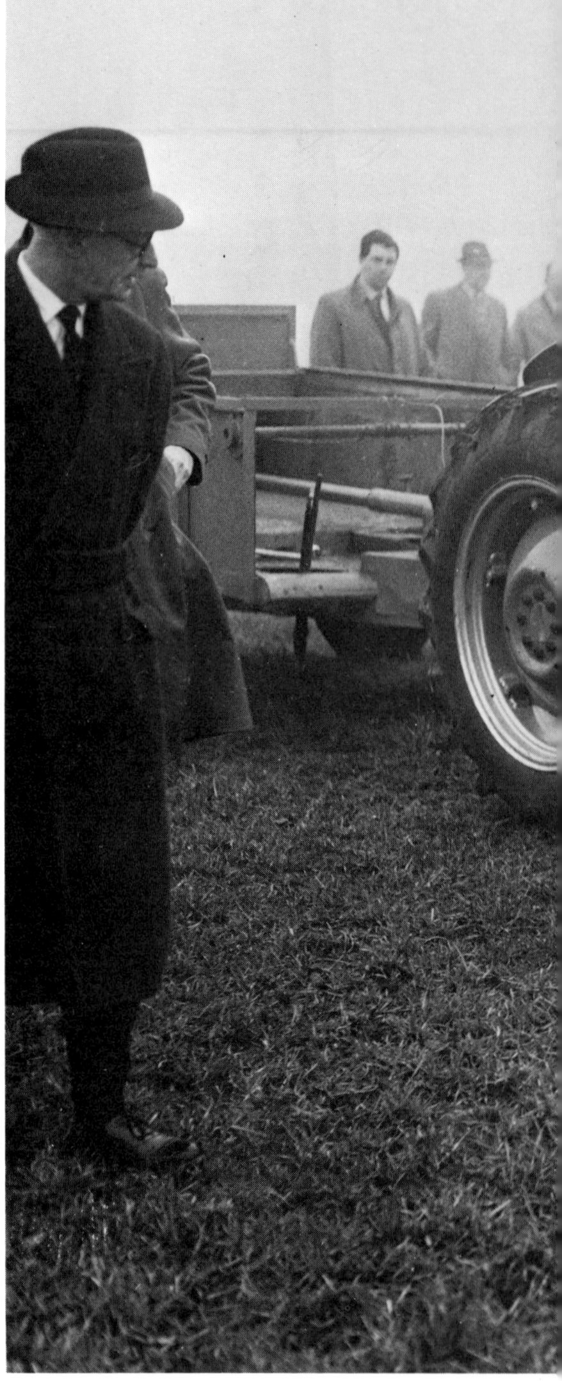

A demonstration of Reading University's Driverless Tractor

PDP 559

system was reliable, they failed to establish the idea commercially. There were criticisms that the driverless tractor was too limited in the range of jobs it could perform and doubts about the economics and safety of the system.

A safety system was built into the tractor to cut out the engine and apply the brakes before the tractor could hit an obstruction. A metal rod carried at the front of the tractor was designed to trigger off the safety cut-out when contact was made with anything solid. This was intended to stop the tractor safely if an object, such as a parked car, was left on the route the tractor was following. It would also provide reasonable protection if, for some reason, the tractor left the guide-wire system and steered at random.

One reason why this and other driverless tractor systems have made little commercial progress is that none of the mechanisms so far devised is a complete substitute for a human operator. The human driver can identify problems, such as a blocked spray nozzle or a soft tyre, and take appropriate action. A driver can use judgement and decide to bale the driest swaths first or to stop the tractor if his trailer-load of bales looks unstable.

Field operations on most farms are not as easy to automate as the repetitive jobs in factories or in highly intensive livestock units. In the foreseeable future, most of the world's farm tractors will have human drivers.

DAVID BROWN 850 AND 950, OLIVER 500 AND 600

The 850 and 950 models were a stage in the development of the David Brown tractor range in the 1950s and 1960s, a period of considerable technical progress when model changes were frequent.

Until 1956, the styling of the principal David Brown tractors had been based on the rounded lines of the VAK series, first produced in 1939 and continued through the Cropmaster and its derivatives. The new, more modern styling, introduced in 1956 with the 900 tractor, survived through a succession of new models until 1965, when the distinctive white and chocolate brown colour scheme was introduced with another styling change.

The 900, with four engine options available, survived less than 3 years before being replaced in 1958 by the first version of the 950. Two engines were offered on the 950, a four-cylinder diesel and a petrol engine, both with a 42.5-hp power rating. The specification of the 950 was changed several times during its 4-year production run as a system of automatic depth control was added to the hydraulics and a dual speed power take-off became available. The diesel engine of the 950 was used in another new model, the 880, which superseded the 950.

With the smaller 850 series the model changes were less complicated. This was a 35-hp diesel, announced in 1960 as an addition to the David Brown range. A petrol version was also available initially, but was discontinued as the demand for diesel power became more widespread. The 850 survived through various changes and developments until 1965, when it was replaced by the 770 with a new three-cylinder diesel engine.

In 1960, some of the 850 and 950 tractors came off the production line at Meltham in Yorkshire with a new look. This was the result of an agreement between David Brown and the Oliver Corporation. Under the terms of the deal, David Brown manufactured some tractors with a green and white paint finish, to be shipped to the USA, where they were marketed as part of the Oliver tractor range.

Advantages of the agreement were increased production volumes for the David Brown factory, while allowing Oliver to rationalise the range of different models they had to manufacture without restricting the choice available to their dealers and customers.

The Oliver version of the David Brown 850 was known as the 500 and the David Brown 950 became the Oliver 600. Apart from the change of colour and name, the tractors were also given a slight restyling to suit the Oliver range, while the mechanical specification remained basically the same.

Under the agreement, which lasted more than 3 years, more than 2000 tractors were despatched from Meltham under the Oliver name.

Above and right. *The David Brown 950 and the Oliver 500, both from the same production line.*

LAMBORGHINI 5C

After World War 2 ended in 1945, most of the countries which had been directly involved in the fighting had large quantities of surplus military equipment available. This was often sold by the Governments concerned and, at a time when European industry was still recovering from the War, the army leftovers provided a valuable source of raw materials.

Ferruccio Lamborghini was an Italian who started a business in this way. Equipment bought from the Government provided some of the materials he needed to build simple farm equipment.

In 1949, he built his first tractor, the start of a successful business which began to grow quickly. At that time Lamborghini had a keen interest in sports cars and these provided an additional outlet for his talent as an engineer.

High performance cars became a part of the Lamborghini product range in 1963 and soon became better known than the tractors. Lamborghini cars offered a combination of speed, luxury and style at prices which made them amongst the most exclusive cars in the world.

Meanwhile the tractors were continuing to sell successfully, including the 5C crawler model which had been one of the first in the Lamborghini range to attract attention outside Italy. The 5C appeared at the 1962 Paris Agricultural Show where it created considerable interest. The feature which drew the crowds was the addition of three wheels with rubber tyres which allowed the tractor to travel on public roads.

The three wheels raised the tracks above the ground. The tracks of the 5C were driven from the rear sprocket and the rubber-tyred wheels were fixed to these sprockets to provide propulsion on the road. The small wheel at the front was simply a caster wheel which allowed the tractor to turn. Steering was by means of the skid effect achieved by using the steering levers, and results could have been quite dramatic if the levers were used too firmly with the tractor travelling at its maximum road speed of 10 mph.

Lamborghini had designed the 5C to suit the Italian requirement for a tracklayer and the tractor was a success. The three-cylinder diesel engine de-

Lamborghini's 5C crawler with its three road wheels in position for the Paris Show.

veloped 39 hp and the tractor was available in a narrow version for vineyard work as well as the standard agricultural model. An unusual feature was the provision of two power take-off shafts, one turning at 2000 rpm in an anti-clockwise direction and the other operating at 570 rpm with clockwise rotation.

In theory, the road wheels were an attractive idea, offering much more versatility than a conventional crawler tractor. Presumably there were some practical difficulties because there is little evidence that the wheels were widely used and Lamborghini later abandoned the idea.

In 1972, the car and tractor interests of the Lamborghini company were split. Lamborghini tractors were taken over by the SAME company, which has continued to develop the Lamborghini range and increase its sales volume.

INDEX

Figures in **bold** refer to colour plates and those in *italic* to black and white illustrations.